Around the World by Stirling Engine

Environmentally friendly Stirling engines,
their applications worldwide
and into space

Brent H. Van Arsdell

Brent Van Arsdell

American Stirling Company
2726 Shelter Island Dr. # 172
San Diego, CA 92106

Designed by: Michael Crumpton
Cover photo/art: Photograph courtesy Kockums
Illustration by Michael Crumpton

Additional copies of this book are
available through the following:
Phone: 760-742-2727
Fax: 858-777-3459
Internet: www.stirlingengine.com
Email: brent@stirlingengine.com

ISBN: 0-9713918-0-7

03 04 05 06 07 • 5 4 3 2 1

American Stirling Order Form

		Quantity	Subtotal
MM-1 Coffee Cup Stirling Engine $179.00	Shipping to the USA $12.00 Elsewhere $16.00		
MM-5 Coffee Cup Stirling Engine Kit $119.00	Shipping to the USA $12.00 Elsewhere $16.00		
MM-6 Stirling Engine $359.00	Shipping to the USA $17.00 Elsewhere $34.00		
Victorian Look Stirling Engine $1479.00	Shipping to the USA $35.00 Elsewhere $89.00		
Smart Stirling Lab Engine $4950.00	Shipping to the USA $65.00 Elsewhere $350.00		
Around the World by Stirling Engine Book $17.95	Shipping to the USA $5.00 Elsewhere $9.00		
	Total		

American Stirling Company
2726 Shelter Island Dr. # 172
San Diego, CA 92106
Phone 760-742-2727 • Fax 858-777-3459
www.stirlingengine.com
email brent@stirlingengine.com

Order Form

Dedication

To Dr. Curtis Hieggelke, without whose
excellent teaching of physics
I never would have become an engineer.

Acknowledgments

I owe many thanks …

to Darryl Phillips, who showed me my first Stirling engine and did the concept for the MM-1 Coffee Cup Stirling engine.

to Kelly Hawthorn O'Connor, who was the first person to bug me about getting the book done.

to Shannon Parsons and Kathy Bazan, who transcribed the interviews.

to my long-time friend and editor, Tim Lale. Tim, your work made this book much more coherent than it would have been otherwise.

to Michael Crumpton, an outstanding artist and industrial designer whose pleasant reminders that I needed to get this book done really helped.

to my parents, Ronald and Jeannette Van Arsdell. Thanks for all the help with my education.

to my friend Kevin Harrington, who helped me put technology in perspective.

to my computer programmer, Jason Tauber, who helped me build a great Web site.

to Richard N. Bolles, whose excellent book, *What Color Is Your Parachute?* helped me move from a career I hated to one that I love.

to Alan B. Johnson, who read my rough drafts and helped me improve this book. Alan, if you ever wanted to switch to editing instead of engineering, I'm sure that you could.

to David J. Gingery, who told me that he could tell by just talking to me that I could write a book. Thanks for the encouragement.

to Lindsay. Your catalog of books convinced me that there is a market for a book like this.

The most difficult part about saying thank you is that I am sure to miss someone important. And so to you, that special someone whom I overlooked in a moment of temporary insanity, I want to say, "I couldn't have done it without you."

Biography

Brent H. Van Arsdell is an expert on Stirling engines. He has been quoted on Stirling engines in the *Boston Globe,* the *Christian Science Monitor,* and the *New York Times*. His television appearances on Stirling engines include appearing on MSNBC ("The News with Brian Williams"), the History Channel ("Modern Marvels: Engines"), and Tech TV ("Tech Live").

His publications on Stirling engines include the segment on Stirling engines in the three-volume library reference book, *Macmillan Encyclopedia of Energy* (2001).

Mr. Van Arsdell is a graduate of the University of Illinois with a Bachelor of Science in aeronautical and astronautical engineering. In addition to being a Stirling engine expert, Brent H. Van Arsdell is a pilot, a technology buff, and a die-hard airplane enthusiast.

Mr. Van Arsdell is available for speaking engagements on Stirling engines. If your group is looking for a speaker, please contact him using the contact information in the back of this book. His current contact information is also available on the American Stirling Website at http://www.stirlingengine.com/contact-us/.

Contents

Chapter 1
Understand Stirling Engines in Ten Minutes or Less

What is a Stirling engine anyway? Stirling engines are a category of engine, just like diesel engines are a category of engine. They are closed-cycle engines, which means that the air (or another gas such as helium) is used over and over again inside the engine. The air inside the engine is never used up and dumped into the environment as dirty exhaust.

Because the air is used over and over again, Stirling engines can be "regenerative." This means that some of the heat used to expand the air in one cycle can be used again to expand the air in the next cycle. Stirling engines do this by moving the air across another material called a regenerator. Regenerators are widely used in industry today and are probably Robert Stirling's most important invention.

When the hot air in a Stirling engine flows over the cool regenerator, some of the heat from the air flows into the regenerator, heating it up. This pre-cools the air before it moves to the cold side of the engine, where it will reject more of its heat and complete the cycle. Compare this process to not having to throw away all of the heat that a conventional internal combustion engine dumps into its exhaust at the end of each cycle.

The inventor of Stirling engines, Robert Stirling (1790-1878), was part preacher, part inventor, and 100 percent Renaissance man. He was a minister of the Church of Scotland who was interested in the physical health and safety of his parishioners in addition to the well-being of their souls. He invented the Stirling engine (he called it an "air engine") because the steam engines of his day would often explode, killing and maiming those who were unlucky enough to be standing close by. Robert Stirling's engines couldn't explode, and they produced more power than the steam engines then in use.

In 1816 Stirling received his first patent for a new type of "air engine." The engines he built, and those that followed, eventually became known as "hot air engines." They continued to be called hot air engines until the 1940s, when other gases such as helium and hydrogen were used as the working fluid. Robert's brother, James Stirling, also played an important role in the development of Stirling engines.

If you'd like to read more about Robert Stirling and the engines that he built, read the biography of Robert Stirling by Robert Sier entitled Reverend Robert Stirling D.D.: A Biography of the Inventor of the Heat Economiser & Stirling Cycle Engine (L A Mair, 1995).

The excellent book by Finkelstein and Organ titled Air Engines: The History, Science, and Reality of the Perfect Engine (American Society of Mechanical Engineers, 2001), has shown that Robert Stirling must have had a much more sophisticated understanding of his engines than is commonly thought. Indeed, according to the simulations done by Professor Organ of Cambridge University, if Robert Stirling had made a mistake as small as selecting the wrong diameter of wire for his engine's regenerator, his initial engine of 1816 would have failed to run at all.

Stirling engines are continuous combustion engines. This can be a tremendous advantage when it comes to building a very clean-running engine. Strike a match and let it burn for a second, then blow it out. When did you see most of the smoke? Most of its smoke was produced not when the match was burning brightly, but when it was being lit and when it had been blown out. It is much easier to make a continuous burning process very clean than it is to make an intermittent burning process, such as those used in most gasoline and diesel engines, clean.

In a Stirling engine, one side of the engine is continuously heated while the other side is continuously cooled. First the air is moved to the hot side, where it is heated and expands, pushing up on a piston. Then the air is moved through the regenerator (where it stores some heat for the next cycle) to the cold side, where it cools off and contracts, pulling down on the piston. The temperature change inside the engine produces the pressure change needed to push on the piston and make the engine run.

Stirling engines are different from the engine in your car in several ways. First, Stirling engines are external combustion engines. This simply means the fire that heats the engine and makes it run does not occur inside the working cylinder of the engine. Most people are familiar with steam engines, which are also external combustion engines. Because the Stirling engine is an external combustion engine, it can be made to run on any fuel or heat source, from burning biomass to solar energy. Of course, the hot side of the engine has to be designed for the chosen heat source or type of fuel.

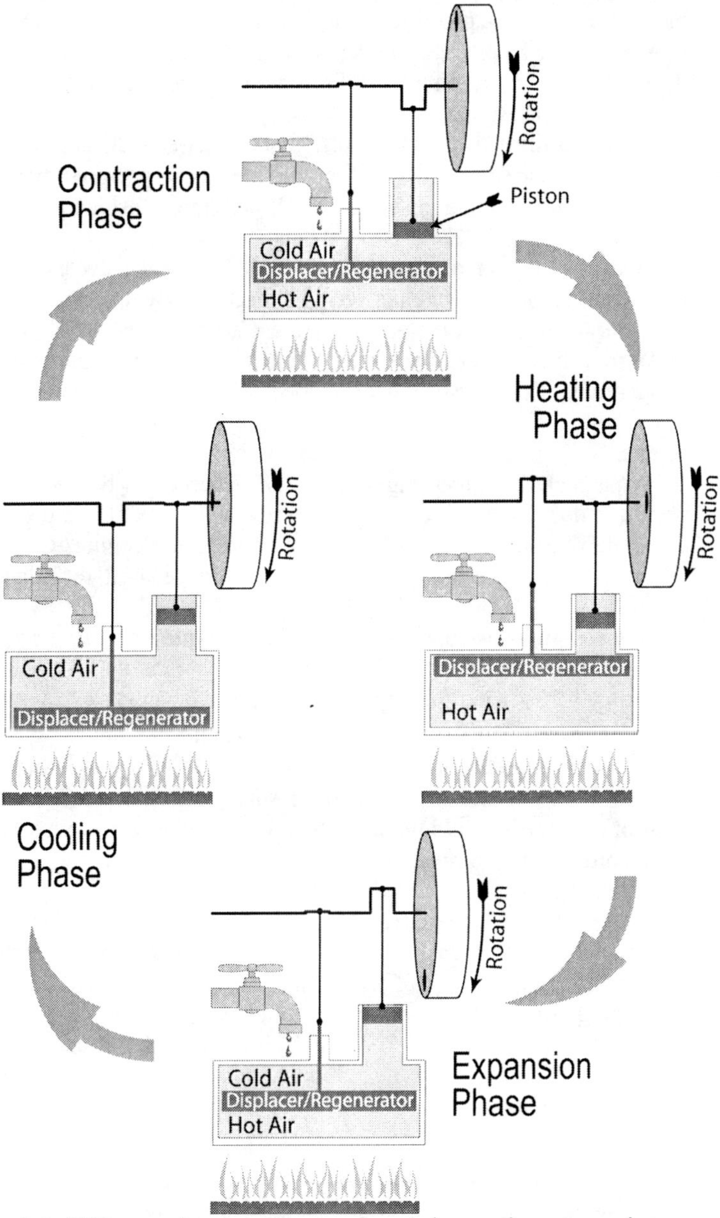

Fig.1-1: Stirling engines work by moving air (or another gas such as helium) between the hot side and the cold side.

Another way Stirling engines differ from car engines is that they have no valves. The pressure changes inside the engine are created by moving the air back and forth between the hot side of the engine and the cold side. Any fuel that will keep the hot side hot can be used to power a Stirling engine.

Fuel is a third factor in the differences between Stirling engines and a typical car engine. Stirling engines have been powered by many kinds of fuel, including gasoline, diesel, propane, sunshine and salad oil.

Because they save some of their heat from each cycle in their regenerators, Stirling engines can be exceedingly efficient. Their efficiency can approximate that of the Carnot cycle, which is the maximum theoretically possible efficiency for any heat engine. Incidentally, Robert Stirling developed and built his engine many years before Sadi Carnot (1796-1832) wrote about the ideal heat engine.[1]

What makes a good engine anyway? Does it do the work you want it to do? Do you like the way it performs its task? Does it satisfy the goals you had in mind when you purchased the engine? Can you afford it? These are the questions that should be used to judge an engine.

Some engine developers don't ask the right questions. I read about a Stirling engine that achieved a record high thermodynamic efficiency while being fueled by sunshine. The developers forgot that efficiency is only important if you have to pay for the fuel.

While it sometimes might be important to get the maximum amount of power out of the smallest solar collector area possible, most of the time it would make sense to design an engine with a large solar collector and a small initial purchase price, since sunshine is free.

Engineers have a specific term for describing the "goodness" of any machine. This term is "figure of merit." The figure of merit is simply the thing that you are trying to optimize when you are developing something. If a low initial purchase price is very important, then the figure of merit is low purchase price. If high thermodynamic efficiency is important, then the figure of merit is thermodynamic efficiency. If long life is important, then the figure of

---

[1] Unfortunately, engine efficiency is a slippery term. It has different meanings to engineers than it does to the general public, and even among engineers there is discussion of thermodynamic efficiency, mechanical efficiency, and electrical efficiency. Sometimes an engineer will claim a very high level of efficiency for an engine, without being very specific about which type of efficiency he or she is referring to.

merit may be the time between overhauls. If you drove to work today, did you ride on an 80-cc motor scooter?

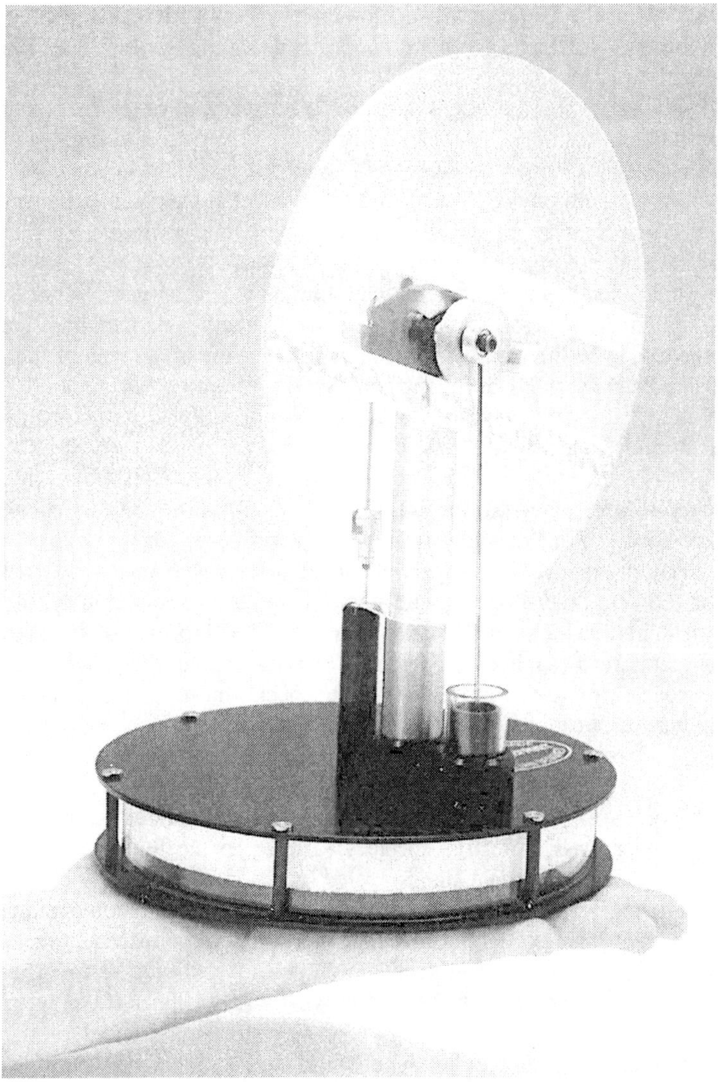

Fig. 1-2: The MM-6 Stirling engine from American Stirling will run on the heat of your warm hand.

If not, then you can be sure that your figure of merit for commuting is something other than thermodynamic efficiency. Perhaps safety or utility was more important to you than miles per gallon.

While Stirling engines can be exceedingly efficient engines, only when they are matched to the right application will they give their owners what they really want. People want not just engines; they want what the engines will do for them. They want lights without pollution, hairdryers without cords, and sailing yachts that run silently.

Power is absolutely essential to life. Every advance in power production from the first water wheels along rivers, to the engines of an airliner on its way to Paris has given food and life to people who would have otherwise died. Today the world is desperately hungry for *clean* sources of power, and for some power needs, Stirling engines are nothing short of the perfect engine.

Let's think for a minute about low-pollution engines. A well-designed continuous combustion Stirling-engine burner has the potential to be the cleanest running engine for any given type of fuel. The advantage that continuous combustion gives to a Stirling engine is that it does not have to do the start-stop portion of combustion many times per second.

What about zero-pollution Stirling engines? Well, there are several different ways of doing this. An interesting, if not practical, zero local pollution engine would be a Stirling engine powered by liquid nitrogen. Industrially available liquid nitrogen would be used for the cold side of the engine while the heat source would be the atmosphere, which, on a typical 72° F (22° C) day provides plenty of temperature difference to operate a Stirling engine very well. This approach might be useful for some special-purpose industrial or military applications.

A more practical and less fanciful way to build a zero-pollution Stirling engine is to operate the Stirling engine on solar power. Solar-powered Stirling engines typically use solar concentrators such as mirrors or other reflective surfaces to heat up the hot side of the Stirling engine. The cold side is cooled by air or water.

Stirling engines have the rather amazing characteristic of being reversible engines. Here's how that works. If you heat one end to operating temperature and keep the other end cool, you can extract mechanical work out of the shaft. But what would happen if you were to connect an electric motor to your Stirling engine and turn the Stirling engine with a motor? You would find that one end would get hot and the other end would get cold. This is how Stirling coolers work.

A motor connected to a properly designed single-stage Stirling cycle cooler can produce temperatures as low as 80° K (-315° F). More complex multi-stage Stirling cryo-coolers can cool to even lower temperatures. These coolers are being built by defense contractors for use in night-vision devices, missile heads, and other applications that need portable cryo-cooling.

Since Stirling engines can be used as coolers or freezers, should your house have a Stirling cycle freezer or refrigerator? In the 1980s, Sun Power Inc. of Athens, Ohio developed Stirling cycle refrigerators. Unfortunately it is extremely difficult to compete with well established Freon cycle refrigeration systems that have been manufactured by the billions. The most valuable role for Stirling cycle coolers today is in cryogenic research situations, where the researcher needs a small, moderately priced method for generating liquid oxygen or liquid nitrogen in a lab. While other methods are used for making liquid air in industrial settings, Stirling cycle coolers are the preferred method for generating cryogenic temperatures on demand on a small scale.

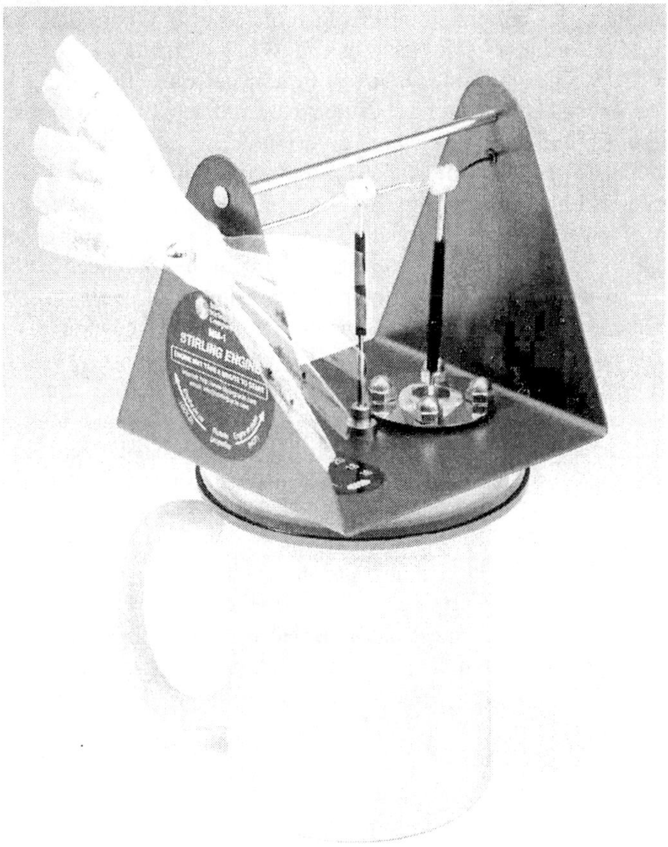

Fig.1-3: The MM-1 Stirling engine from American Stirling will run on the heat of a cup of hot water or ice cubes.

Of course there are some advantages of Stirling cycle refrigeration that are tempting. For instance, Stirling cycle refrigerators use environmentally friendly helium instead of CFCs. Also they can be modulated to produce extremely low power efficient refrigeration instead of cycling on and off the way vapor cycle refrigerators do. Stirling engines can be built that will run on extremely small temperature differences. Figure 1-1 shows the model MM-1 Stirling engine from American Stirling Company™. This engine will run on a temperature difference of about 40° F (22° C). If you think that's impressive, you should take a moment to examine the model MM-6 (Figure 1-2), also from American Stirling Company™, that runs on a temperature difference of 15° F (8° C) and will run on the heat from your warm hand in a 72° F (22° C) room.

Since the world is full of millions of waste heat sources that produce small temperature differences, it's tempting to think that these engines can be scaled up to produce commercially viable amounts of power. Unfortunately, that approach leads to excessively large Stirling engines. Building a Stirling engine that will produce a lot of power requires a relatively large temperature difference and typically cannot be done with small temperature differences.

How many different types of Stirling engines are there? The different types of Stirling engines can be roughly compared to the different types of internal combustion engines. In other words, there are many more than you might think. With internal combustion engines, you have V-8's, V-6's, straight 6's, straight 8's, straight 12's, radials, rotaries, and so on. Stirling engines have about as many different configurations, but there are three common configuration types.

The first configuration type is known as the Alpha and has two pistons that move the air back and forth between the hot side and cold side.

The Beta configuration has a displacer in the cylinder, in the same cylinder as the piston. The Gamma configuration has a piston and a displacer in separate cylinders.

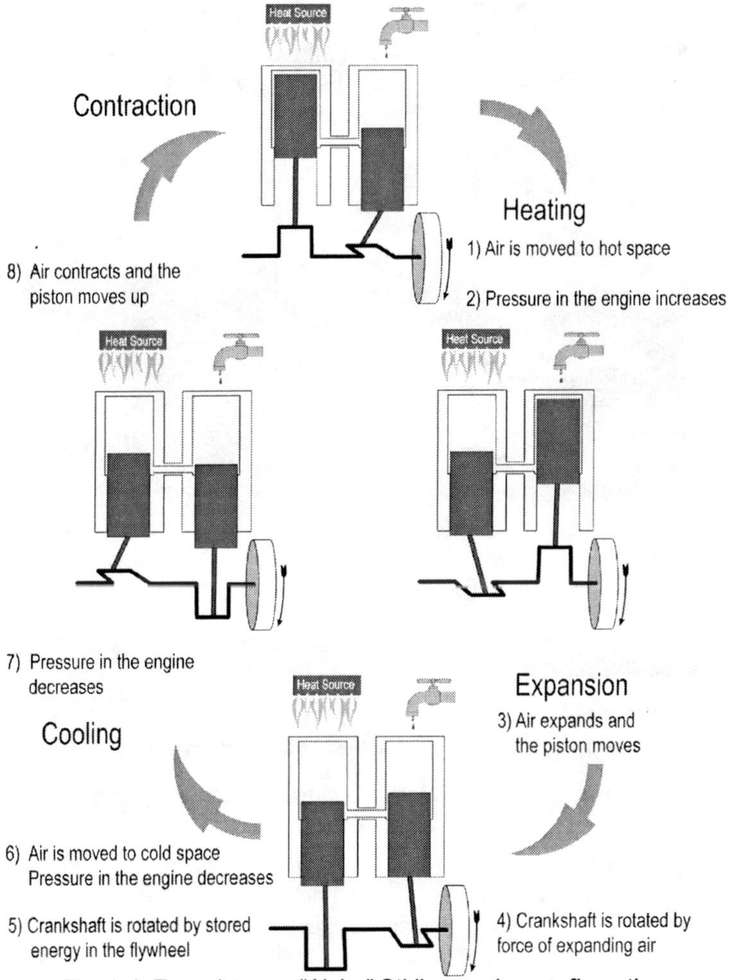

Fig. 1-4: Two-piston or "Alpha" Stirling engine configuration.

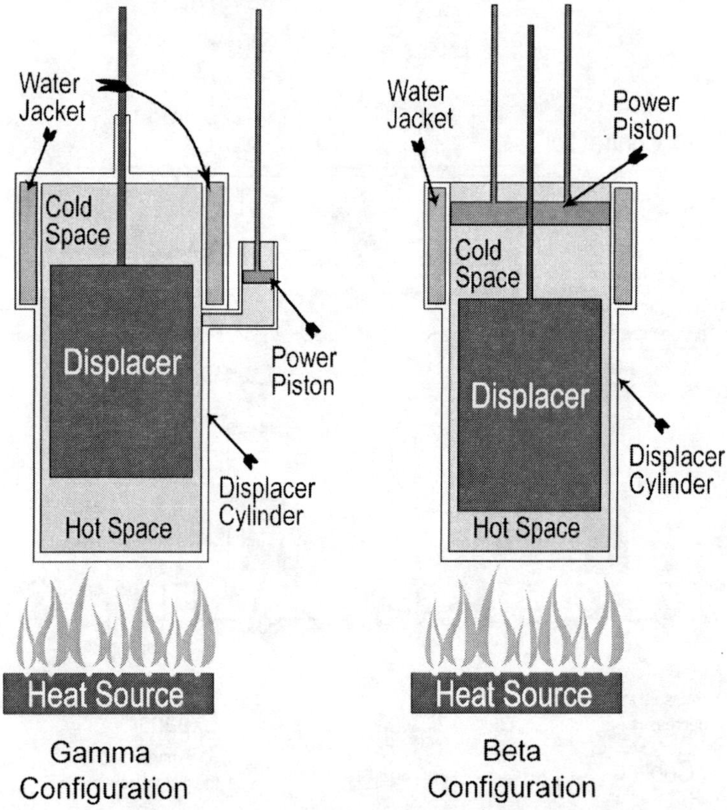

Fig 1-5: The Gamma and Beta configurations of Stirling engines.

There are hundreds of different mechanical arrangements that can achieve the Stirling cycle. Each configuration of Stirling engine has an application in which it excels. Some configurations are right for submarines while others are right for solar power. And some are right for providing combined heat and power in homes.

So now that you know about many of the different types of clean, quiet, and efficient Stirling engines, you may ask yourself, "If Stirling engines are so great, why don't I have one in my car?"

Chapter 2
If Stirling Engines Are so Great, Why Don't I Have One in My Car?

In the 1930s, Philips Electronics, headquartered in The Netherlands, was looking for an engine to power its radios. An employee of Philips came across a Stirling engine at a flea market. After purchasing the engine and using it, he realized that Stirling engines could provide the silent and efficient power source that Philips wanted to run their radios.

By the time Philips perfected the engine for use as a generator to power their radios, improvements in electronics had dramatically reduced the requirements for electrical power, eliminating the need for a separate generator. This left Philips with a rather well-developed small engine. Philips engineers realized that the perfect application of Stirling engines was in boats.

They put one of their small research engines in a row boat and motored around the Dutch canals with no apparent source of propulsion. The excellent cooling provided by the water kept the cold side cold while the engines were typically fueled by fossil fuels. Philips developed larger marine engines up to 90 horsepower at one point, powering a small motor yacht.

Marine applications were an excellent match for Stirling engines, providing silence that people liked, and the cooling water available for the Stirling engine dramatically simplified the installation. In the 1970s, as the oil crisis emerged, a tremendous amount of government research money became available to develop alternate sources of power. Most of this money was targeted at the automobile industry. Philips was contacted, and its marine engines were put into service in cars, buses, and other road vehicles.

Another reason that Stirling engines do not work well in cars is that they are inherently slow-starting engines. Automobile drivers are used to being able to get into their car, start their engine, and immediately be able to burn rubber across an intersection. Stirling engines start more slowly because it takes some time for the hot side to get up to operating temperature. They take about the same amount of time to start as a jet engine or diesel engine would take. Slow starting is a fundamental feature of Stirling engines and there is almost nothing that can be done about this. Using a Stirling engine in an application where instant starting is required is probably a bad idea.

Fig.2-1: Stirling engines are not a very good choice of engines in cars.

It is very easy to regulate the power output of typical internal combustion engines. A simple throttle mechanism provides fast, efficient, and reliable regulation of power across a wide range of power outputs. Stirling engines really want to run at a constant power setting all day long. Fortunately it is easy to make a Stirling engine power-regulation device that regulates the power efficiently but slowly. Just imagine turning the burner up and then turning the burner down.

Fig. 2-2: Engine compartment of Stirling powered AMC P-40 Spirit.

It is also easy to design a Stirling engine power-regulation mechanism that is extremely fast but very inefficient. Imagine opening a valve that shuttles the gas between the hot side of the engine and the cold side, bypassing the power piston. (Jim Symanski's engine, shown in chapter 10, uses a mechanism like this.)

But it's not easy to develop a Stirling engine that can regulate its power level both efficiently and quickly. When the major automobile companies wanted to develop a Stirling engine for use in cars in the 1970s and 1980s, they ended up with extremely complex mechanisms that pumped helium or hydrogen from storage to the working spaces of the engine in order to regulate the power output of the Stirling engine. It's not a good idea to use Stirling engines in any application in which you must be able to instantly change the power settings from very high power to very low power both quickly and efficiently.

Stirling engines were widely used from the 1870s until about the 1920s as domestic water pumps, power sources for sewing machines, and for other household and business purposes. But the antique Stirling engines were built of materials such as cast iron. When cast iron is heated up to the average temperature of the flame, it oxidizes very quickly, making early Stirling engines unreliable. Modern Stirling engines can be built with high-temperature steels that resist oxidation.

Internal combustion engines became the engine of choice for powering automobiles before Stirling engines did, and they are much better suited for that application than Stirling engines. While internal combustion engines can use the same flame temperatures that Stirling engines use, their extremely high temperature flames only occur for a very short period of time, providing cooling time between cycles. In an internal combustion engine, the hot portion of the engine never gets up to the average temperature of the flame. This allows the use of cheap materials such as aluminum and cast iron in these engines. On the other hand, in a Stirling engine the hot portion of the engine *does* rise to the average temperature of the flame. This requires using a small amount of a high-temperature stainless-steel alloy. With modern manufacturing techniques, this is a small cost of the complete engine but it should not be overlooked.

It is worthwhile to note that while Robert Stirling was a preacher, a scientist, and an engineer, he wasn't a salesman. He also didn't write a book on how to properly design Stirling engines for other applications. The only documentation that he left for others were his groundbreaking patents.

Internal combustion engines can be easily scaled. If you have access to a photocopier with a reducing or enlarging feature, you can scale an internal combustion engine. It's almost that simple. Stirling engines can be scaled as well; however, the rules for scaling are not linear at all. Stirling engine scaling rules include things like surface-area-to-volume ratios, Mach number, and the Stirling engine number. If you do fourteen out of fifteen things right in designing an internal combustion engine, you will still get an engine that works, but if you do fourteen out of fifteen things right in a Stirling engine and you get the thermodynamics wrong, you have just built an expensive piece of modern art. You will want to enjoy what it looks like on your coffee table, because it won't be an engine.

It's worth noting that the Stirling engines built by Robert Stirling and his brother, James, in the 1840s and 1850s were significantly better engines than those that were widely produced by his imitators from about 1870 up to 1920. The later copies almost never had regenerators and had significantly lower efficiency than the engines that Robert and James Stirling built.

Stirling engines should be used today for applications in which their silence, smoothness, and multi-fuel ability allow them to accomplish the job where nothing else will work. They are not simply a different machine for turning gasoline into power. Stirling engines can be the quietest engines in the world across a wide range of power outputs. Engines as small as the MM-6 (see Figure 1-2) and as large as full-power submarine engines from Kockums (see the next chapter and the cover photo) are totally silent engines. Space engines are another area in which the Stirling cycle excels. Engines for powering spacecraft need to produce an extremely low level of vibration in order to avoid damaging delicate sensors on the spacecraft.

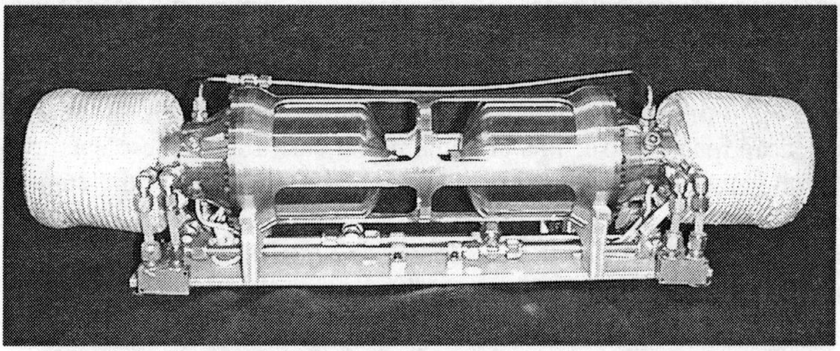

Fig. 2-3: This dual Stirling engine spacecraft power package is manufactured by Stirling Technology Company of Kennewick, Washington. It can run for years without maintenance.

Stirling engines can be extremely low–vibration engines because their combustion is smooth and external. There are no explosions within a Stirling engine. Also the power output of the Stirling engine is extremely smooth. They can be the smoothest of any engine imaginable. Since silence and low vibration go together, it's only natural that the most prominent modern use of Stirling engines is in the submarine. In the next chapter, we will discover why the Stirling-powered submarine may be the most dangerous thing in the ocean to any navy that doesn't have one.

Chapter 3
The Most Dangerous Thing in the Ocean

An interview with Mr. Christer Bratt, Project Manager for Stirling Air Independent Propulsion at Kockums, a Swedish defense contractor, conducted by the author.

Q: Why is silence so important in a submarine?

Bratt: In a submarine, the enemy is always looking for you, and since sound is the primary method for detecting a submarine, it's much more accurate to say that the enemy is always listening for you, so it's very important that no one can hear you. And if they can't hear you they probably can't find you.

Q: Explain what the propulsion system is like in most submarines?

Bratt: Most submarines use an electric motor to propel the submarine, although the nuclear submarines of the United States Navy have a direct couple to the steam turbines. Most submarines, including all the submarines that we build, are "diesel electric" submarines.

In a diesel electric submarine, the diesel engine powers a generator that runs the motor. In a sense, it is very similar to the diesel electric locomotive in a train. The main difference is that in diesel electric submarines there will also be many tons of lead acid batteries that can power the electric motor when the submarine is submerged below snorkel depth.

Q: How is a Stirling engine integrated into a diesel electric submarine?

Bratt: The Stirling is an "add-on system" on the submarines we have today. So you still have a diesel engine. You still have an electric propulsion motor. You still have a hundred tons of lead acid batteries. But you also have a Stirling engine that powers a generator. When the submarine is on the surface it can charge its batteries with the diesel engine, and when it is submerged, it can maintain a full and operational charge of its batteries for much longer than it could if it didn't have the Stirling engine.

Q: What is the approximate power output of the electric motors of one of your typical submarines, such as, say, the Nacken Class sub?

Bratt: It's approximately one megawatt (~1,300 hp) for both the diesel engines and the electric motor.

Fig. 3-1: This Gotland Class Submarine has Four Stirling engines that provide auxiliary power.

Q: It's my understanding that your Stirling engines are much smaller engines than the main propulsion motors of your submarines.

Bratt: That is correct. We have a 75 kilowatt unit (100 hp), and there are between two and four of these units on these submarines.

It's not used to charge the batteries. That's enough power to run the submarine at very low speed, and to run the life-support systems and the electronics. It dramatically increases the endurance time of the submarine.

We used to say, if a conventional submarine with batteries could stay submerged for a couple of days, with a Stirling it could stay down a couple of weeks.

Q: Two days to two weeks. That's a pretty good dramatic difference. What fuel does the engine use?

Bratt: It uses the same supply of diesel fuel as the diesel cycle engine, and it uses cryogenic liquid oxygen as the oxidant.

Q: What do you run out of first, diesel fuel or liquid oxygen?

Bratt: You run out of oxygen first because there is a limit on the size of the oxygen tanks. You use about 4:1 oxygen to diesel. The Stirling system with the oxygen tank is very simple to have. It's sort of like a battery that you charge with liquid oxygen. You also get an energy content which is perhaps four to five times more than that of a lead acid battery.

Q: If you were to start a new submarine design with a clean piece of paper, might you design a submarine that was Stirling/electric with no diesel engine at all and fewer batteries?

Bratt: Yes. That is what we are looking into for the future for submarine designs.

Q: Stirling machines can be configured as cryo-coolers. Is your Stirling system also a cryogenic generator that could make its own liquid oxygen when it is on the surface?

Bratt: No, when you build a Stirling cooler, you tend to do it differently than you would a power producing Stirling engine. Our system uses ordinary industrial grade liquid oxygen; it's the same liquid oxygen that you might use for welding or in a shipyard. And it's very easy to refuel. It takes less than an hour to refuel. It can even be done at sea.

Q: I've been told that the combustion occurs at very high pressure in your engines. Why is this?

Bratt: We need the high pressure in order to get rid of the excess products of combustion. Otherwise we would have to pump it out, and that could make noise and it would consume power. If you have high pressure in the combustion chamber, you can just have a valve and have it go out into the water. Most of what goes out is carbon dioxide.

Fig. 3-2: Two to four of these Stirling engines provide dramatically improved performance in Kockums submarines.

There are devices aboard to dissolve the carbon dioxide in the sea water so that it won't make noisy bubbles. The pressure for the high-pressure combustion is created by the oxygen tank so we keep the oxygen tank at high pressure. We then feed the oxygen into the combustion chamber at the high pressure and then inject the diesel fuel in so the combustion takes place. The combustion pressure allows diving to a depth of more than 200 meters.

Q: Does the engine recycle part of its exhaust?

Bratt: It uses an internal recirculation of the exhaust in order to keep the combustion flame temperature down. When you burn pure oxygen and diesel fuel, you get a very extremely high temperature. Think of how a cutting torch works to burn steel and you get a good idea of how hot the combustion can be.

Q: Now, was Kockums a licensee of the Philips Stirling engines or did you develop this engine independently?

Bratt: We started as a Philips licensee in 1968, but this engine was developed independently.

Q: What is the basic configuration of the 75 kilowatt units that you use?

Bratt: It's a V-4 engine with double acting piston. There are two balance shafts to keep the vibration very low.

Q: So how much vibration does the engine produce?

Bratt: Oh, it's very low because of its two balancing shafts, and the engine is mounted in a special way to avoid all vibration. So it is very, very quiet.

Q: Someone told me that you could set a coin on edge and it wouldn't vibrate off.

Q: Yes, that is true.

Q: If I were standing in the room next to it and it were running, would I hear it?

Bratt: Not in the submarine. It's built in. It's encapsulated, you might say. When you have guests we sometimes ask the guests to tell if the engine is running or not when they are standing next to it. When the engine is installed in its enclosure it's not possible to hear if it's running or not.

Q: What is the physical size of the 75 kilowatt unit?

Bratt: The engine itself is about 1.3 meters high (4.26 ft) with a width and length of 700 millimeters (2.29 ft). If you were to look at just the engine by itself it is about the same size and weight as an equivalent power diesel engine. Of course that assumes that the engine was running on air. The high-pressure combustion system used in the submarine makes the entire package somewhat larger than a diesel engine.

Q: How long will the engine last?

Bratt: Of course that depends on the application, but the life of a submarine is something like 30 years. Of course, you have to do the maintenance comparable to what you might do to a diesel engine over that time.

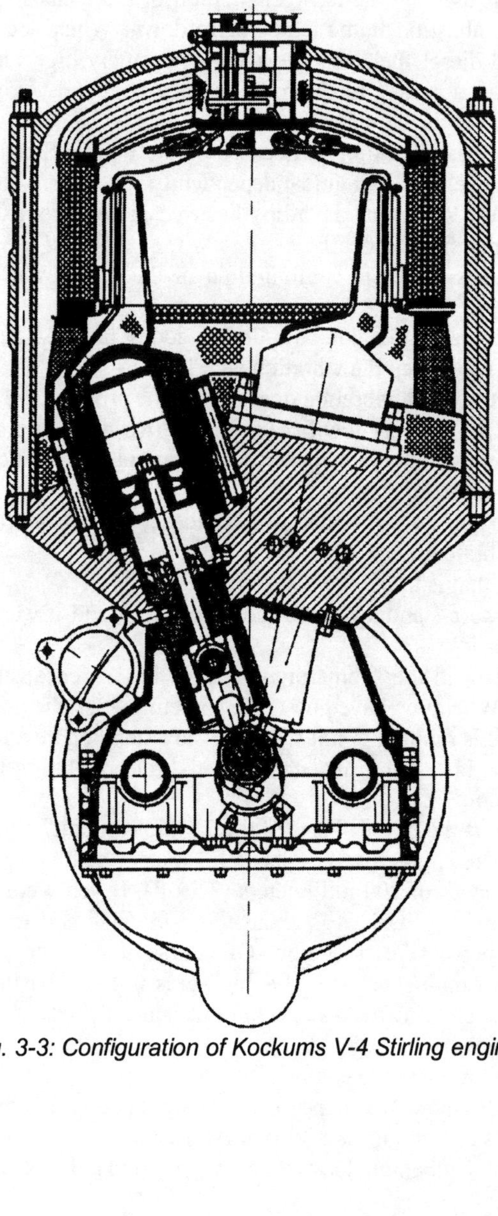

Fig. 3-3: Configuration of Kockums V-4 Stirling engine.

Fig. 3-4. This Stirling engine goes from very quiet to absolutely silent when the panels are closed.

Q: Your company makes some very high quality, very quiet submarines. How much harder is it to detect a submarine with this system than one without it?

Bratt: I'm not a military officer myself so it's difficult for me to give good answers, but this should give you an idea. Without a Stirling power system like ours, a submarine captain might put up the snorkel, run the diesel engine and recharge the batteries perhaps every day or every other day. Something like that. When a diesel electric submarine is at snorkel depth recharging its batteries, it can be detected from airplanes and by other methods fairly easily.

With our Stirling power system a captain can stay down at operating depth, and if he's deep he's also being very stealthy. With our system this stealth time can be extended from days to weeks.

Q: Was Kockums the first company to put a Stirling engine in a submarine?

Bratt: Yes, and to my knowledge no other company has put a Stirling engine in a submarine. We now have many years experience with Stirling engines. They are very reliable.

Q: If you were to design a new submarine for a country that could choose any type of power plant, would a Stirling power plant be the best choice?

Bratt: That's a theoretical question. Nuclear power plants have a certain minimum size, so they tend to size the entire submarine. For example, if you look at Seawolf, the most expensive submarine, an American attack submarine, they are 8,000 tons. I don't think you can build a nuclear submarine under 1,500 tons. Now an 8,000-ton submarine is very useful in the deep ocean, but in places where you have shallow water, you can't use such a big submarine there. Then there is price. I think a nuclear submarine costs about twenty times more than a small Stirling submarine.

Q: Has anyone ever built a mini-sub or a research sub powered by a Stirling engine?

Bratt: Yes. The very first installation that we made was in the French research sub called the *Saga.* It's about 500 tons of displacement. It was originally built by Jacques Cousteau with underwater scientists. We delivered two Stirling engines for that submarine. This was back in 1984-85, I would say.

Q: And were they also 75 kilowatt (100 hp) engines?

Bratt: Yes. Yes. I would say it was a very early version of it. They did some of their early tests in the Mediterrean. It would dive down to 600 meters, with the Stirling engines running.

Q: One of the most interesting uses of Stirling engines I have seen proposed is the idea that luxury submarines could be built. Could a private submarine yacht be built and be powered by a Stirling engine? Is that possible?

Bratt: Technically, it is quite possible. It just costs money to have such a system.

Q: Has Kockums considered developing a 1 megawatt (1300 hp) Stirling engine to power the entire submarine?

Bratt: Yes, but those are long-term plans. It won't be this year.

In the next chapter you will learn how Stirling engines may power the world's first luxury submarines.

Chapter 4
The Ultimate Yacht
A Stirling Powered Luxury Submarine

An interview with L. Bruce Jones, President, U.S. Submarines Inc., Ft. Lauderdale, Florida, by the author.

Q: Your company makes and markets private submarines, mostly tourist submarines. Now you are promoting the idea of a luxury private submarine, and you are looking for a lead customer. What kind of person do you think will be the first to own the "Ultimate Yacht"?

Jones: Who will the first buyer be? That's hard to say. We've talked to a lot of very high-end businessmen. We've talked to one or two heads of state, and some Middle East royal family members. That's about all I can tell you. It'll be someone with money for sure. I can say that the strongest interest seems to come from people who have a scuba diving or yachting background. I think we will end up building the first one on spec.

We have not yet sold any of our bigger submarine designs. We have a $25-million design, called the *Seattle 1000*. It's about 113 feet [34.5 meters] long and displaces about 380 tons. We've also done considerable design work on a submarine called the *Phoenix 1000*, which is 213 feet [65 meters] long and displaces 1,500 tons, has about 5,000 square feet (464 square meters) of interior space.

Beyond that, we have a project where we have been designing the world's first submarine cruise ship. It will carry about seventy-two passengers on two- and three-day submerged voyages. That's called the *Poseidon 1000*. It's about 286 feet [86 meters] long and displaces about 2,300 tons.

Q: How does a $25-million submarine compare in price with the world's mega yachts?

Jones: There's a dozen yachts under construction that are over 100 meters (328 feet) long. A minimum price for a yacht in that size range is $110 or $120 million and they go up to $200 million. Paul Allen, one of the founders of Microsoft, is building a 400-plus foot yacht now at Lurssen. That yacht is expected to cost $200 million or so. So $25 million is sort of a drop in the bucket for some of these people who are into conspicuous consumption.

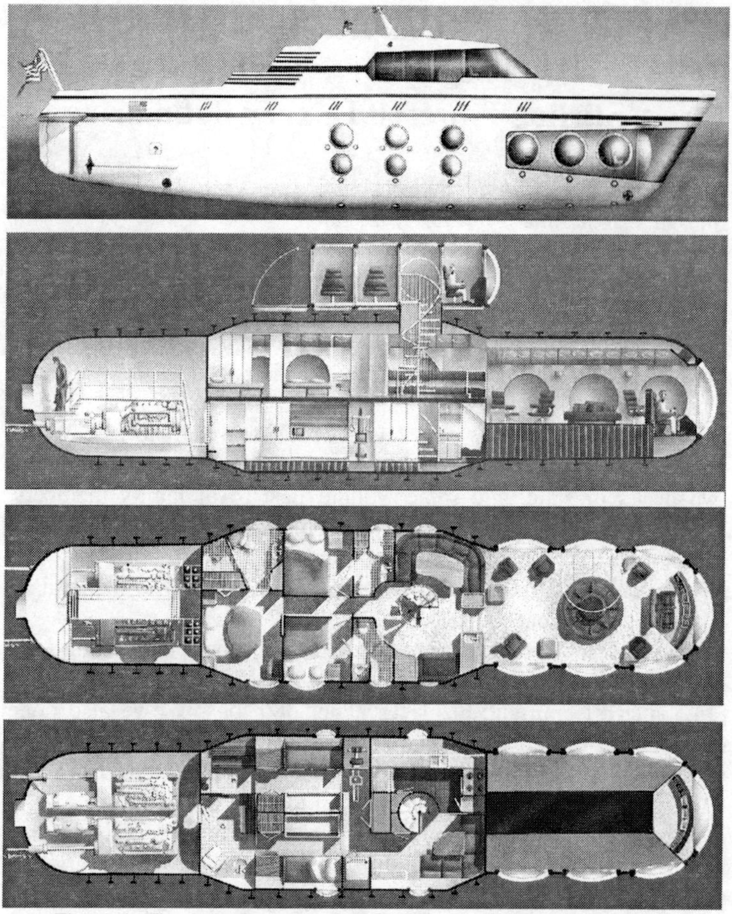

Fig. 4-1: The luxury submarine Seattle has been designed by U.S. Submarines.

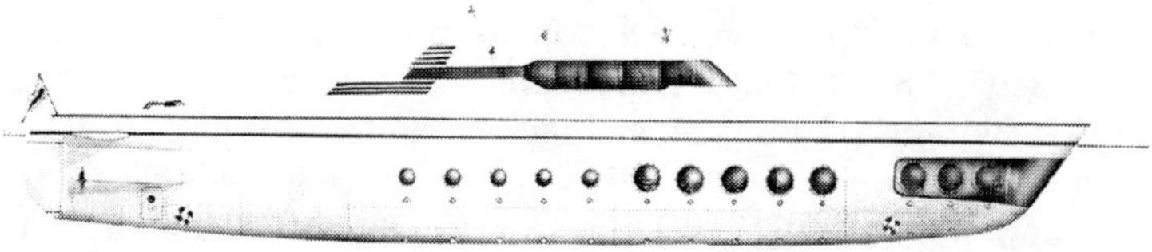

Fig. 4-2: With more interior space than two average American houses combined, the Phoenix 1000 has plenty of room for luxury.

Q: So it's clear that the economics potential is there for your idea of luxury submarines to work.

Jones: That's true. The difficulty in finding a luxury submarine buyer is that most people don't have any submarine experience. And what experience they do have is military submarines, so they tend to think of submarines as cramped and ugly. I think that until we build a couple of them and get them out there as demonstration units, it will be an uphill battle for luxury submarines.

Q: There have been heads of state with submarine backgrounds. Perhaps a person like former U.S. President Jimmy Carter might be a lead customer?

Jones: Maybe, but maybe not. There's a huge difference between military submarines and luxury submarines. I think it's far more likely to find a guy who's a diver and a yacht aficionado who will make the transition, rather than an ex-submariner.

Q: Has anyone ever converted a military submarine to private yacht status?

Jones: No, and that wouldn't really make sense. But there have been some military submarines which have been used as museums. There's four or five of those in the U.S.

Military submarines are designed as weapons platforms, and as a result, they have a lot of redundant systems. You wouldn't be able to expect to operate a military submarine at all cost effectively. You'd have to have a crew of thirty-five to forty guys at a minimum, and then the operating costs would be astronomical. That's why it's never been done. That and the fact that military submarines, just by their very nature, have high tensile strength steel pressure hulls that are extremely difficult to work with.

Q: Do you think that there should be a "Navy One," that might be the rough equivalent of Air Force One?

Jones: Well, the president of the United States has lots of options such as underground bunkers and Air Force One. It might make sense for the head of state of some other country.

Q: Tell me about the civilian submarines currently in use.

Jones: There are forty tourist submarines operating worldwide. They carry passengers on short-term, one-hour trips to depths of up to about 300 feet [91 meters]. The industry carries approximately two million passengers per year.

Fig. 4-3: The tourist submarine Voyager cruises in front of a city skyline. Tourist submarines seat up to sixty passengers.

Q: That's more than I expected! Are there other categories of private submarines?

Jones: There are about thirty active research subs worldwide. These are generally owned by research organizations and governments. There are a few small diesel electric subs that are privately owned for personal use. There are a small few small all-electric submersibles that are launched from private yachts by their owners.

Fig. 4-4: Tourist submarines carry more than two million passengers per year.

Q: You mentioned a submarine that is designed for cruise tourism. Tell me more about how that might work.

Jones: The plan is to pick a good base of operations where the submarine can be based on a long-term basis. This would be analogous to a cruise-ship experience where passengers would have their own private state room, each with a six-foot [two-meter] diameter view port. You'd have underwater lounges and dining areas. We'd have naturalists lecture on board. Cruises would probably last up to three days. It would be quite an expensive proposition: in 2003 dollars we're looking at $4,600 per ticket. The submarine itself would be pretty expensive. For a seventy-two-passenger cruise ship it would be about a $140-150 million project.

Q: Could Stirling engines be an important part of any of these submarines?

Jones: The *Seattle*, the *Phoenix*, and the cruise ship are all very good applications for an air independent propulsion system such a Stirling engine. Here's why.

Typically with manned submersibles such as tourist subs and smaller submarines, they are so small in terms of interior volume, that people tend to enjoy trips for shorter periods of time. Battery power is perfectly adequate for what they need to do. But when you start designing bigger submarines that are more comfortable or have mission profiles that would require dives of longer duration, then you really have to look at something beyond conventional lead acid batteries to provide power to propel the submarine and power all the lights, hot water, and all the hotel electrical loads while you are underwater.

Q: What are the underwater power requirements for the motors on submarines like the *Seattle* or the *Phoenix* class?

Jones: It depends on what the owner wants to do. If the owner is happy to simply stay submerged and move around pretty slowly underwater for a few days or to stay parked on the bottom for two weeks, if they were really into doing some observation or something like that, conventional battery power in submarines the size of the *Phoenix* or the *Seattle* is a fine way to accomplish that.

But if you start talking about the extended sub-sea travel, if you want to go several hundred miles or more underwater, before you surface, then you need another form of propulsion. So there are a number of different air independent propulsion systems, AIP systems for short, that have been utilized in submarines in the past, and Stirling engines are one of them.

In the case of something like the *Poseidon*, the submarine cruise ship, conventional battery power just isn't going to do it if you want to stay underwater for two or three days at a time and travel several hundred miles. That's where closed cycle Stirling technology is applicable. The technology has matured to the point where it is certainly viable.

Q: Are there any factors that limit how well a submarine designer can use a Stirling engine?

Jones: Yes, you need to store considerable quantities of liquid oxygen. Of course that's true for other air independent propulsion systems such as closed cycle diesel too. Here's what happens. As you use the liquid oxygen, you have to compensate for the weight of the oxygen you have burned by bringing salt water on board and storing it, somewhere other than in the liquid oxygen tank.

Of course there are two parts to the fuel, but the problem of burning off diesel fuel [a Stirling cycle engine would also use diesel fuel] is easy to solve. What happens is that the diesel tanks are outside the pressure hull. They are bag tanks. They are vented to the sea water with baffles, etc. so the process is automatic. Now there is a difference in density between sea water and diesel, but it's not so much so it's easy to deal with.

Q: Are extended submerged trips with a Stirling-powered submarine possible?

Jones: We've been working on the design of a civil submarine for British Petroleum that would travel around the world submerged in 2,500 mile legs. It would most likely be powered by a Stirling engine. The liquid oxygen carrying requirements for a mission of that duration will be pretty significant.

Q: Is it possible? Could it really be done?

Jones: Oh, yes! It's absolutely possible. It will absolutely work. We have a design to do it. The ratio of liquid oxygen to diesel fuel is 2.8:1, so while the oxygen requirements are significant they can be managed. It's too expensive a project for them to undertake at this point, but I think some day it will be built.

Q: Where does it make sense to operate a private luxury submarine?

Jones: You can operate it anywhere where you would use a normal yacht. Obviously, the better the water quality, the more ideal the experience. People have different ideas about these things. Some people want to be in clear water or dive in a shallow cove. Some people want to dive deep to see what they can find. Other people are hunting for wrecks and want to tow a magnetometer behind the sub. Everybody's got a different idea of what they will be used for. The caveat is that you can't operate submarines in heavily trafficked areas. You have to be pretty careful when coming to the surface.

Q: I grew up near Chicago, where the U-505, a captured German WWII submarine, is on display at the Museum of Science and Industry. I would say that everyone in Chicago who ever thinks of a submarine thinks of the U-505. Tell us a little bit more about the differences between luxury submarines and military submarines.

Jones: Military submarines are designed as weapon platforms; they are made for a very specific purpose. For instance, you can never put view ports in the sides of a military submarine because you've got ballast tanks and you've got all sorts of other paraphernalia. Military submarines are designed to travel at significant speeds almost all the time, while tourist submarines and luxury submarines are capable of hovering in a water column and maneuvering in all three directions with bow and stern thrusters and vertical thrusters. Tourist and luxury submarines are designed to get in close to something and see it with pinpoint maneuvering capability.

Another difference between military and luxury submarines is the complex engineering involved in designing acrylic view ports that meet safety standards. The way those view ports interact with the hull and the port forgings is very complex. It requires a lot of engineering talent to be able to solve those problems, and they are a much different set of problems than the problems a designer of a military submarines faces.

Q: People who have had a chance to tour through a World War II or perhaps a current military submarine have noticed how terribly cramped they are inside. How much room would a luxury submarine have inside?

Jones: They are very comfortable. There are two essential characteristics that make them far more comfortable than military submarines. One is that you've got the large view ports everywhere so you get a lot of ocean light streaming in, and since you have an excellent view of the ocean around you, you get a wonderful ambiance. The second issue is that in civil submarines, we don't require any pressure-proof bulkheads or any internal supports through the interior spaces of the submarine. Military submarines have those things because they are literally expecting people to depth-charge them and shoot at them. That's not very likely in a civilian submarine. Once you build that pressure hull, you can walk from one end to the other without any obstacles, if that's how you want to set it up. Bulkheads can be put up for privacy or for anything the individual client and the interior designer have in mind.

Q: Can the interior space really be open from front to back of the submarine?

Jones: Exactly. You wouldn't normally do it that way but it could be done that way if you wanted to. For example in a big tourist submarine that seats from 48 to 66 people, all the passengers are in one large room. It's sort of like being in a theater for a good movie with a good crowd.

If you look at a 747 or any other wide-body jet, you'll have an idea of the interior space that is possible. Of course a luxury submarine would have two levels, one above the other. They can be very spacious. In the *Phoenix* model there is over 5,000 square feet (465 square meters) of space available.

In the next chapter you will learn how Stirling engines are keeping the drinks cold and the guests warm on some of the world's richest yachts.

Chapter 5
How Stirling Engines Silently Keep the Drinks Cold and the Guests Warm on Some of the World's Richest Yachts

An interview with Don Clucas, technical director, Whisper Tech Ltd., Christchurch, New Zealand, by the author.

Q: All new technologies are expensive when they are first introduced. What types of pioneering customers have you found who are willing to pay for your Stirling engines?

Clucas: Our main market right now is the luxury yacht market. Our engines go into 40-60 foot (12-18 meter) sailing yachts and motor yachts. Other applications are remote homes, research stations, and the occasional lighthouse.

Fig. 5-1: WhisperGen Stirling engine, shown here between a refrigerator and a countertop, provides both heat and power for remote and city locations.

Q: The people who are choosing your Stirling engines have other choices for power generation. Why are they deciding to use your Stirling engines?

Clucas: We have a fully integrated system. When they board the yacht they can turn the unit on, and it will monitor the level of charge in your batteries and also monitor the temperature in your rooms. It will start and stop as needed to maintain a comfortable temperature or charge the batteries on the yacht. Our system is also very low maintenance. You don't have to make messy oil changes like you do with diesel generators. Our system is nearly silent. You don't have the thumping of a diesel engine every time you want to run your microwave.

With the WhisperGen you get your domestic heating, you get your hot water supply plus you get all your electricity requirements.

The owner doesn't even have to know that it is a Stirling engine. All they need to do is to put fuel into it and make sure the cooling system is working. Our automated system does the rest.

Fig. 5-2: WhisperGen engines being readied for customers.

Q: Does the WhisperGen heat a yacht with forced air or hot water?

Clucas: It heats the yacht with hot water at about 70° C (160° F). We have a temperature control which you can see the temperature of the water that comes from the unit. Hot water heat is standard in yachts. It's also the standard heating method for heating homes in Europe and the northeast part of the United States. The Europeans call it "hydronic" heating.

Q: How much power do current versions of the WhisperGen put out?

Clucas: The battery-charging DC units we are selling at the moment are rated at 750 watts. They usually will put out between 800 to 850 watts. The residential AC unit is rated at 850 watts and peaks at 1,200 watts.

Q: And what fuels does the engine burn?

Clucas: In the marine market, we are currently running the WhisperGen on diesel but it can also run on kerosene. We have a unit operating down at the South Pole, which runs on JP8, which is a military aviation jet fuel. There's a range of fuels that it can run on. Any fuel similar to automotive diesel should work. Some people are even testing what they call bio "diesel" fuels, but perhaps we should really call these fuels "bio-Stirling."

Q: Tell me about your different models.

Clucas: We have a model for residential co-generation that has a natural gas or propane burner. We just produced thirty engines with AC power output. Those units are going into homes in the U.K. with natural gas burners. We also make a DC or battery-charger model with a natural gas or propane burner. This would be perfect for off-grid houses, as would the diesel- or oil-fired unit.

Q: Tell me a little bit about the history of this engine.

Clucas: Back in 1989, Peter Lynn and I proposed a project to the University of Canterbury to develop a commercially viable Stirling engine. I did my Ph.D. project in mechanical engineering on the prototype of this machine. I could see that there was a market, especially in the yacht industry, for a quiet Stirling engine, and also the technologies were right for building an engine for that market. It also seemed like a suitable challenge for my Ph.D. project.

In 1993, I presented a paper at the Stirling Engine Conference in Eindhoven, Netherlands. At the same time, I was discussing residential co-generation applications with an electricity supplier here in New Zealand called South Power. The company is now called Orion. At that time, they were very interested in residential co-generation, but they could also see the benefit of remote power systems. So we set up a company called Whisper Tech Ltd., they invested in the company, and I moved away from the university to become technical director of Whisper Tech.

The first thing we did at Whisper Tech was to produce 20 prototype systems which went out onto the various yachts and remote homes. We learned an awful lot from that. It's essential to get your engine out of the lab.

In about 1996, we started working with Gasunie in the Netherlands, a very large gas supply company, to work on residential co-generation systems. We produced the first AC prototypes for Gasunie, based on the WhisperGen with an induction alternator which would plug directly into the grid. We produced a couple of prototypes for them. A little bit later, in 1998, we started working with EA Technology in the United Kingdom. They are an electricity research organization, and they were proposing projects for operating residential co-generation projects.

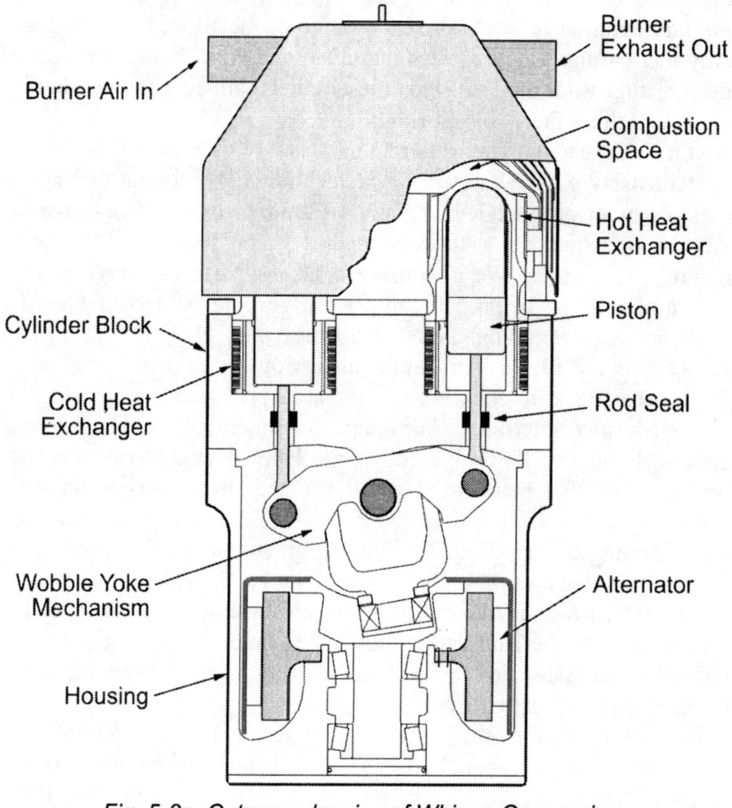

Fig. 5-3a: Cutaway drawing of WhisperGen engine.

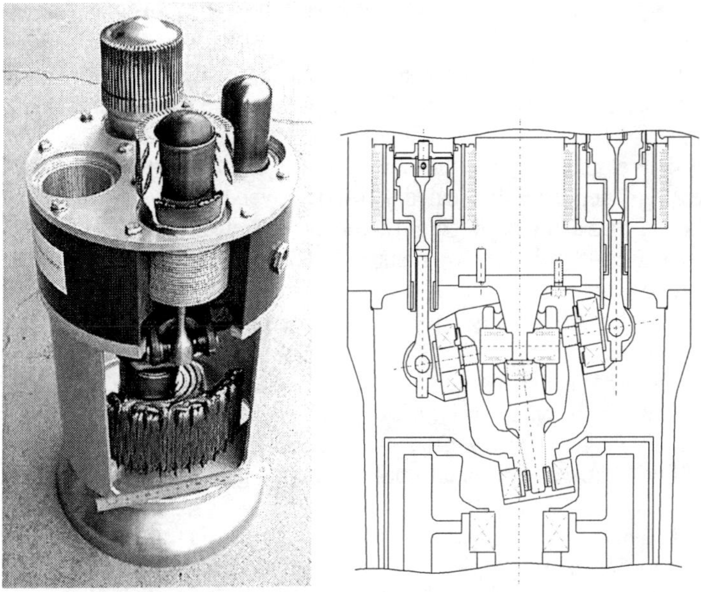

Fig. 5-3b: Cutaway drawing of WhisperGen engine and a disassembled engine to show detail.

Q: Give me a summary of how many engines have been produced and in what types of applications they are being used in.

Clucas: We have delivered more than 300 of the DC systems for the yacht market, and that market seems to be really growing. Most of them are in sailing yachts or motor yachts. There's probably about 50 or so in remote land-based applications. For example, we have two systems at the South Pole for the upcoming winter season. Many of these engines are in remote homes.

We have the atmospheric research organization for the America's Cup using our system to power remote monitoring systems. We've produced about 100 DC gas systems that have gone up to Japan. We're probably getting close to 100 AC gas systems for residential co-generation systems. At the moment, there are about 30 of the latest version, which has been approved for sale within the UK. They are going into residential homes at the moment.

Q: Give me the brief overview of the basic design of the engine.

Clucas: It's a four-cylinder, double-acting machine. The cylinders are grouped around the central axis of the engine, and there is a wobble mechanism that links the four pistons to a rotating alternator. The alternator is hermetically sealed into the unit. Sealing the alternator avoids any problems with dynamic, high-pressure seals which have traditionally been a problem in Stirling engines.

The engine uses nitrogen as the working fluid. The main reason for using nitrogen is that it is a large molecule, which is easy to seal and also it's much easier to obtain than helium. So when you commission an engine in far-off places, there is a pretty good chance of getting the nitrogen for doing the charge of the engine.

The engine is grease lubricated so it's got all its rolling element bearings permanently sealed. The DC version has a permanent magnet alternator producing 12- or 24-volt electricity. We plan on producing 48-volt DC versions as well. The AC machine for the European market uses a 230-volt induction alternator. Both engine models get double duty out of the alternator by also using it to start the engine.

Q: What is the engine's exhaust temperature?

Clucas: Probably about 80 to 90° C (176 to 194° F). The exhaust heat exchanger is a very efficient condensing heat exchanger.

Q: What types of metal does the engine use in the hot section of the engine?

Clucas: High-temperature stainless steel. It's not a significant cost portion of the engine. We're about to move to an even better manufacturing technique so the cost of the metal will be even less significant in the future.

Q: How long does it take to install this engine in a yacht?

Clucas: If you've got a new installation it will probably be about a day's work to install it. For an existing yacht, it might take two days to a week depending on how long it takes to route the hot water lines.

Q: Tell me more about some of the interesting places where your engines are being used.

Clucas: The University of New South Wales had one running last year through the winter season in a small, bubble-shaped sort of vehicle that sits out on the ice at the South Pole. The first year's test went so well that they ordered two more units for an even more remote location this year. Our engine will power an infrared telescope power supply through the Antarctic winter. The telescope is at 3,500 meters altitude (11,500 feet). It's probably setting an altitude record for land-based Stirling engines.

This is the first power supply they have found which can meet the demands because it's running at an outdoor air temperature of -80° C (-112° F).

Q: Are you saying that the engine was turned on in a certain month and then shut off six months later? Did it run continuously for six months?

Clucas: The initial one had a few glitches unrelated to the core engine, like the exhaust freezing up, but after we solved that problem the engine operated unmanned for most of the season. The next two engines have just been installed, and these are set to operate unmanned for the entire Antarctic winter season.

Fig. 5-4: A WhisperGen Stirling engine powers this research station at the South Pole. It runs through the Antarctic winter entirely unattended.

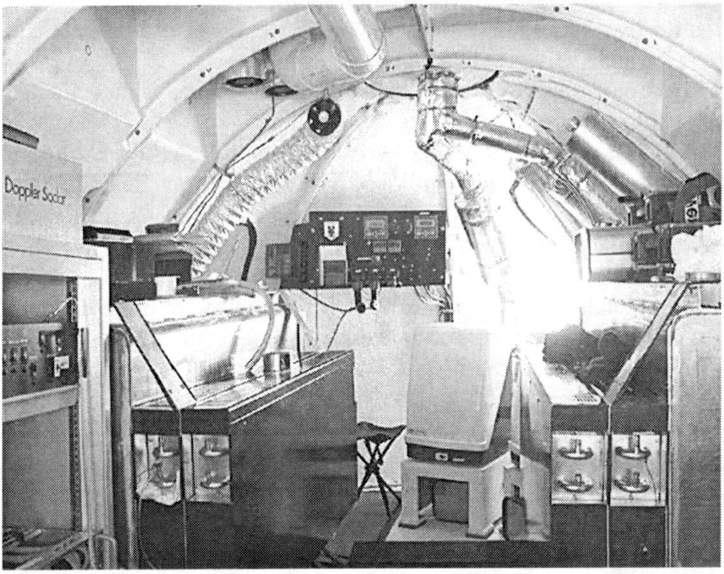

Fig. 5-5: This WhisperGen Stirling engine (center of photo next to the window) powers an infrared telescope and other scientific equipment.

Q: What type of people do you think should be calling you up as potential customers?

Clucas: People with remote homes should call us. Anyone whose situation means that it will be very expensive to get electricity into their house. If your house is going to be off the grid you should call us. People with yachts should call us, because when they get to a harbor they don't want to have their big engines running overnight, generating electricity. It's much more convenient to turn on a WhisperGen to supply heat and power.

There's also a lot of interest from people with research stations. Some of our engines are being used by the U.S. Border Patrol, which uses them to power cameras that monitor the border between the U.S.A. and Mexico.

Some electricity utilities that want to do distributed generation have called us. We are getting inquiries from electricity companies and gas companies. Sometimes gas companies want to get into electricity, and sometimes electricity or combined electricity-gas companies simply want to add new capacity.

Q: Should environmentalists be interested in the WhisperGen?

Clucas: Absolutely! We've gotten a lot of inquiries from departments of conservation, whether it's remote camping huts or remote monitoring systems—they can use one of our gas systems. You don't have any concern about diesel or things like that because if you have a leak, it just evaporates. Our system is also very quiet and clean burning, which reduces environmental impact.

Q: Could this engine be used to power houses on a wide scale in a large city?

Clucas: Absolutely! When electricity is generated by burning a fuel at a central power station, over half of the energy content of the fuel is immediately wasted as heat. That's why power stations often have huge cooling towers. Of the 50 percent (or less) that is left, a percentage gets lost every time it goes through a transformer or down a wire. Usually much less than 40% of the energy from the fuel that was burned is actually used in the home as electricity. Most of the fuel is simply wasted, with a lot of extra CO_2 being produced that doesn't benefit anyone. Combined heat and power plants (CHP) make use of the heat that would usually be wasted. These small power plants can be located in homes or businesses where the electricity is used so that transmission losses are negligible.

Until now, combined heat and power has been confined to large industrial users. Whisper Tech has taken the CHP to another level by producing a WhisperGen for use in homes. The system replaces the normal gas-fired furnace (or hot-water boiler). It only operates when heating is required, so while it heats your home and your domestic hot water, it produces electricity which is fed back into the house. This means about 90 percent of the fuel's energy content is used. The result is a dramatic reduction of CO_2 emissions for the electricity generated and a much more efficient use of fuel resources.

The WhisperGen CHP system also benefits electric utilities by providing an additional electricity supply within their network. This reduces the need for costly network upgrades and new power plants. As the units are installed into homes, the expensive approval processes for building large power stations is avoided. The capital cost for adding new generating capacity is also reduced as the generator is supplied as part of the home heating system at a price very close to that of a traditional home heating system.

In the next chapter you will learn about a larger Stirling engine that is silently making power on Flamingo Avenue in Las Vegas.

Chapter 6
Turning Sunshine Into Power in Las Vegas

Interview with John Zolan, vice-president of marketing and sales, STM Power in Ann Arbor, Michigan, by the author.

Q: I was in Las Vegas and I saw a Stirling engine on Flamingo Avenue that had your company's name on it. It's a test program with the University of Nevada Las Vegas and the local power utility. Tell me a little about that engine and what it's doing.

Zolan: Right. It's our 25-kilowatt Stirling engine operating on an array of mirrors that are focusing or concentrating the sunlight on the end of the engine. As you know, since the Stirling engine is an external combustion engine, it can operate on any source of heat.

Q: What kind of people should be interested in this system after it progresses beyond the beta test stage?

Zolan: It's the same market as the people who would be interested in photovoltaics (solar cells) or other renewable resources, whether those are wind power or what have you. In other words, anyone who is interested in generating electrical power with no emissions would be interested in a system like this.

Q: Is the engine on Flamingo Avenue set up to operate at night with another fuel source?

Zolan: Yes. There are several different ways that these engines can be set up. One of our engines is located in an area where they generate more electricity during the day than they can use, so they use the extra power during the day to produce hydrogen, and the engine burns hydrogen at night. We also have a unit which burns landfill gas at night.

The engine can also burn natural gas at night. That gives you a lot of flexibility in what the machine can do.

It also leverages the asset. Once you've paid the money to purchase the hardware, if you then used it only eight hours or nine hours a day, that isn't nearly as economical in terms of recovering your investment as when you can run it around the clock. Of course people use electricity around the clock, so the ability to use the investment twenty-four hours a day makes sense for a lot of reasons.

Q: How clean is this power that is being produced?

Zolan: When it's running in the solar mode, it's not producing any emissions at all, so it's as clean as photovoltaics or wind power. When it's burning fuel then you should really be comparing it to micro

Fig. 6-1: The author standing in front of a solar Stirling engine array on Flamingo Avenue in Las Vegas, Nevada. This engine was made by STM Power.

turbines and internal combustion engines. It is as clean or cleaner that all of the above. It's significantly cleaner than a diesel engine or a gasoline engine running on natural gas.

Q: Why is it cleaner than a diesel or a gasoline engine?

Zolan: We have to look at the various classifications of pollutants. There's the nitrous oxide, NO_x and carbon monoxide and partially

Fig. 6-2: This solar Stirling system built by STM Power is being operated by a utility near Phoenix, Arizona.

burned hydrocarbons, etc. When you're trying to burn something in an internal combustion engine, you're injecting fuel into a cylinder and lighting it off as an explosion. You have a very hot, very intense, very short duration flame front. The production of NO_x, for example, is tied directly to the maximum temperatures that are achieved while oxidizing the atmospheric nitrogen. With our external combustion Stirling engine, when we are burning fuel in our engine we burn it in one continuous process. The result is combustion that looks a lot more like the heater or hot water heater or in your furnace rather than an explosion.

We have very good combustion control, with a continuous combustion process at lower pressure and a longer dwell time in the combustion zone than other technologies. This results in very complete combustion and very low carbon monoxide. Because we can control the temperature, we do not have a very high temperature spike in the flame front. This lowers the amount of NO_x that we produce as well. So the Stirling combustion system in our engines is actually capable of meeting the most stringent emissions requirements for the California Air Resources Board.

Q: What other environmental benefits are there?

Zolan: We can not only burn clean fuels cleanly, we can burn dirty fuels cleanly. Our engine can burn fuels like methane that's coming off a digester at a waste water plant or at a landfill. These aren't very clean fuels, and most other types of engines don't run well on them. Methane gas that has about 26 times the potency molecule for molecule as a greenhouse gas than carbon dioxide does, so we are able to burn a fuel that otherwise would have been a strong pollutant. So there is an additional environmental benefit too.

Dairy operations, hog farms, and any other agricultural operations that have significant waste treatment problems are also prime customers. If you take the manure from these animals and put it in an anaerobic digester, you can produce methane plus a residue. The residue that is left behind is a pretty good fertilizer, and the methane that is produced can be burned in an STM Stirling engine to produce electricity.

Q: You mentioned that the working fluid in your engine is hydrogen. Ever since the Hindenburg, some people have been concerned about the safety of hydrogen. Can you comment on the safety of hydrogen as a working fluid?

Zolan: Well, there's roughly a half an ounce of hydrogen in the engine. That is a very small amount of hydrogen. And hydrogen has some other safety advantages compared to other things that people are familiar with. Remember that if the hydrogen were to leak out it will quickly rise upwards and away from people. If you were to compare the half an ounce of hydrogen in our engine to the propane tank mounted next to their barbecue, I don't think they need to worry about the hydrogen. If the propane were to leak, it's going to be heavier than air, and it can find its way down your basement steps to the pilot light in your hot water heater. Hydrogen is a very active and light gas, obviously, and it's going to dissipate very rapidly. So from a safety standpoint, compared to propane, hydrogen is actually quite safe. Remember that we aren't burning hydrogen; there is just half an ounce of it which is continually reused in the engine as its working fluid.

Fig. 6-3: This is how STM engine looks when set up to burn landfill gas or other gases.

Fig. 6-4: Eight Stirling engines from STM power burn landfill gas at this site in Michigan.

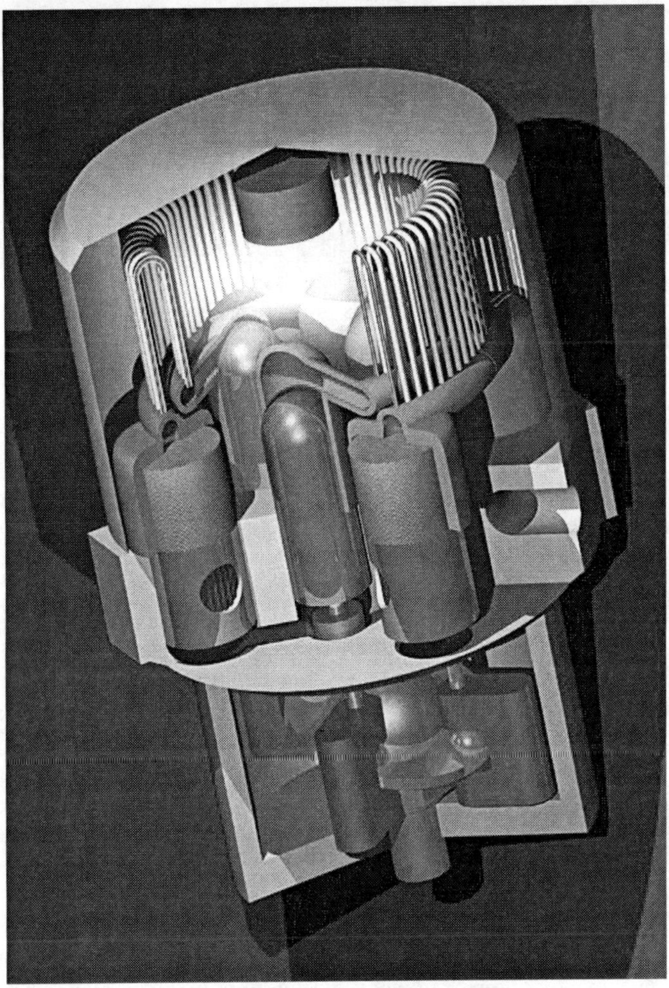

Fig. 6-5: CAD rendering of STM Stirling engine.

Q: Now one of the difficulties with using hydrogen has been that since it's such a small molecule, it will eventually migrate through steel walls and gradually leak out of the engine. How do you deal with that?

Zolan: Yes, that's absolutely true. In fact, hydrogen will even very slowly migrate through solid metal castings. So there is a minor amount of leakage of hydrogen from the unit. We're talking about molecules here, not anything that would be a safety concern.

Eventually the pressure will start to decrease. When the hydrogen pressure in the cylinder decreases, so does the efficiency of the engine. We actually use a device that takes water and a small amount of electricity and produces oxygen and hydrogen. This unit produces the hydrogen at a high enough pressure so that it injects itself automatically into the engine. We continually monitor the pressure of hydrogen in the engine, and when it drops below a certain level, the hydrogen replenishment unit turns on and converts a few drops of water to hydrogen and pushes it into the engine, so we're constantly topping it off.

Q: So people don't have to have a hydrogen tank to keep the engine full of hydrogen?

Zolan: No, all they need is to refill the water bottle. We're talking about a few liters of water every few months. It's not a lot of water, and it's not a lot of trouble.

Q: How many parts does it have, say, compared to an internal combustion engine of a similar power output?

Zolan: Less than half. It's a closed cycle engine. Because the working gas is continuously being recycled inside the engine, we have no valves, there's no valve train, there's no camshaft, there's no timing gears, or ignition system. Because it's a continuous burn process, all that stuff goes away.

Q: So your parts count goes down?

Zolan: It goes down significantly, and of course the number of parts is a cost driver and it's also a factor that you need to take into account with regard to reliability. If the part isn't there it can't break.

Q: So how long do you expect your engines to last?

Zolan: Compared to, say, an automotive derivative industrial IC engine in this size range, it's probably two to three times the expected life span. We're anticipating approximately forty thousand hours of operation from this engine.

Q: And how long will the engine go between maintenance intervals?

Zolan: Other than adding a little bit of water to the hydrogen replenishment system, the engine should only need maintenance about once a year. Because of the fact that it's an external combustion engine, products of combustion never come into contact with any of the moving parts. They don't come into contact with any of the lubricants, so the lubricants never get contaminated. In fact, all the lubricants are all on the cold side of the engine, and since the lubricants never even get hot, the oil lasts a very long time!

Q: What power outputs do you currently offer?

Zolan: Our commercial offerings will be engines that produce 55 kilowatts of electricity. We also have designs on the drawing board to extend our size range to 100 kilowatts and then ultimately to 200 kilowatts.

Q: What do you think the primary applications will be?

Zolan: They will predominately be used in two applications. One will be fuel-fired units, burning a wide variety of both liquid and gaseous fuels, both conventional and renewable. The other type of application will be waste heat recovery applications, in which the heat from an industrial process or an external source can be run through the engine and used directly to produce electrical power. So if you've got an incinerator, a thermal oxidizer, a kiln, a furnace, or a smelting operation that's exhausting high temperature heat, then we can help recover that heat and convert it to electricity.

Q: What temperature would those heat streams need to be?

Zolan: They would need to be optimally in the range of 700 to 1,000° C (1,300 to 1,800° F).

Q: What type of person or organization should be most interested in calling you?

Zolan: Well, our engines have such broad appeal that it's a little bit difficult to narrow down. In general we're looking for a customer who needs electricity or one who wants to reduce their electrical bill or their energy bill. We have to keep in mind that the unit not only produces electricity, but you can also use the unit's waste heat to produce hot water. You can do that very simply and economically. So if you have an application in which you need electricity at a reasonable price and you can use hot water for space heating or in some industrial process, then you would be interested in our Stirling cycle engine.

Of course the operating cost of the engine gets reduced if you happen to have a free source of fuel; if you're close to a landfill or if you're using agricultural digesters or a waste water treatment plant. Having that type of fuel available is a plus. Biomass is another possible fuel. Wood chips or pellets, corn, sugar cane, and a wide variety of agricultural byproducts can be burned externally to the STM Stirling engine, and the resulting heat can be used to generate electricity.

Because we have such flexibility in the combustion, we can also burn types of liquid fuels such as used motor oil. So a commodity that for most people is a waste disposal problem will in our case be a fuel.

Q: Compare your Stirling engine to other technologies for making clean power such as photovoltaics. Everyone understands photo cells. How would you compare your Stirling engines to photovoltaics?

Zolan: When you compare our solar application to photovoltaics, our unit is probably half the cost and twice as efficient. Our engine will take less space and is less costly.

Our engines can also be operated at night, while burning a fuel, while the solar cells only work when the sun is shining.

Q: Compare your Stirling engine technology for recovering waste heat to a steam plant used for the same purpose.

Zolan: The difference is, when you're dealing with a steam plant, you're dealing with a system that's relatively high in cost and complexity. First, you have to have a water makeup system to produce water that is clean enough to use in the boilers. Then you have a complex plumbing system with large volumes of steam going through it. Next you have to have boiler water heaters with large heat exchangers, a cooling tower, and a condenser. There is a lot of claptrap to make a steam plant work, and because of the complexity of the fundamental steam cycle. It really doesn't make a lot of sense for most customers until you're talking about several megawatts of output.

If you have a smaller waste stream that you want to convert to electricity, then our standard 55-kilowatt product is a great way to start. When we look at installing multiple Stirling engines our technology makes sense in the 50-500 kilowatt range and possibly higher, depending on the application. Of course as soon as you start dealing with the steam above a certain pressure in most jurisdictions you're required by law to have an engineer or certified steam operators on site. Of course we avoid those costs with our engines as well.

Q: How many of your units are in use right now?

Zolan: We currently are in our beta test phase. We've built 50 units, and they are located in places ranging from Flamingo Avenue in Las Vegas to Oregon to Taiwan. We have units in agricultural applications with digesters, some in waste water treatment plants, and some in landfills burning landfill methane. Several are burning propane, and others are using waste heat directly.

Q: Prices for products will vary, but what do you expect your engines to cost?

Zolan: I anticipate that the price for the fuel-burning engine will initially be under $1,000 per kilowatt. The solar unit is obviously going to be more expensive because in addition to the engine that's at the focal point of a solar collection array, you've got to purchase the array. Since the array that concentrates the sunlight has to track the sun and concentrate the sunlight with some precision, it will add to the cost of a solar system.

Of course the $1,000 per kW price is with our initial production. Our company is using an automotive production model, and is located in Michigan because we're close to the automotive industry with its well-developed base of high-volume suppliers. Since we have fewer parts than an internal combustion engine, we have some very well thought out but aggressive plans to lower our costs.

Clare Van Ausdal, a senior service technician at STM Power, answered some additional questions.

Q: Does your engine run on both solar power and fuels on the same engine? In other words, can it make the transition from burning fuels at night to running on sunshine during the day?

Zolan: Yes, that's correct. It's a hybrid system. We're also working on a system right now that can run solar and gas at the same time in case a sunny day becomes a partly cloudy day.

Q: Ever since the Hindenburg disaster in 1937 some people have been concerned about the safety of hydrogen. How would you describe the energy potential stored in the hydrogen in your engine compared to the energy potential in a gallon of gasoline?

Zolan: It's much less than a gallon of gasoline. In fact, it takes 400 cubic feet of hydrogen to equal the same BTU value as gasoline. This hydrogen in this engine probably has the explosive capability of 1/100 of a gallon of gas.

Q: So if you're not worried about driving your car down the road, you shouldn't be worried about being next to this engine.

Zolan: Correct. As a matter of fact, the chance of the hydrogen harming you on this engine would be almost zero.

In the next chapter you will learn about a Stirling engine that women are guaranteed to want!

Chapter 7
A Stirling Engine That Women Are Guaranteed to Want

People like convenience. Anything that does something better, or faster, or easier has the potential to be a winner in the marketplace. So what about the possibility of building a cordless hair dryer? This would appeal to the millions of people who would prefer the additional convenience of being able to dry their hair after a shower without tripping over the blow dryer's cord. There would also be additional safety benefits, as you would be removing an electrical appliance from an environment in which there is usually lots of water and the potential for electrocution.

In addition to the safety and convenience benefits of using this type of hair dryer at home, travelers would enjoy the convenience of not needing to pack any cumbersome electrical voltage adapters.

Cordless hair curling irons and cordless hair styling combs are already on the market. These devices are usually powered by a small propane canister that is custom-built for the application. A Stirling engine–powered hair dryer could fit very well into existing marketing channels. In the book *Air Engines: The History, Science, and Reality of the Perfect Engine,* Allan J. Organ, a recently retired engineering professor at Cambridge University, was the first to write about the possibility of Stirling-powered hair dryers. I spoke with Professor Organ by phone at his office in Cambridge.

Q: In your book *Air Engines* you talked about the possibility of building a Stirling-powered hair dryer. Please tell me how you became interested in this idea.

Organ: It was an idea that was brought to me by Cambridge Consultants. Cambridge Consultants is an organization that takes ideas from companies and commercializes them. They don't manufacture things themselves; they help other companies take the initiative and get the production going themselves. One of their big commercial successes has been in high-speed printing. So, they came to me asking me to carry out a design study of the power unit of the hair dryer. I don't know if you're familiar with the Braun hair styling comb, which is gas heated and completely portable. They are now selling like hot cakes. I bought one as a background for the study, and they're actually quite remarkable things. They have engineered them to be completely safe even though they have a flame inside them! It's a combustion process which appears to be assisted by surface catalysis.[2] There is a tiny hole in the unit, and you can just barely see the glow of the burning gas inside. You buy the fuel cylinders separately, and judging by the number of spare cylinders in stock, they are quite widely used.

There is a general trend towards making domestic appliances portable, and the Stirling powered hair dryer is part of that trend. Another reason for doing this project is that other hair dryers that claimed to be portable have very poor performance; they don't dry the hair well, and they don't produce an adequate blast of hot air. The Stirling engine powered hair dryer was expected to be comparable in performance to a salon-type model.

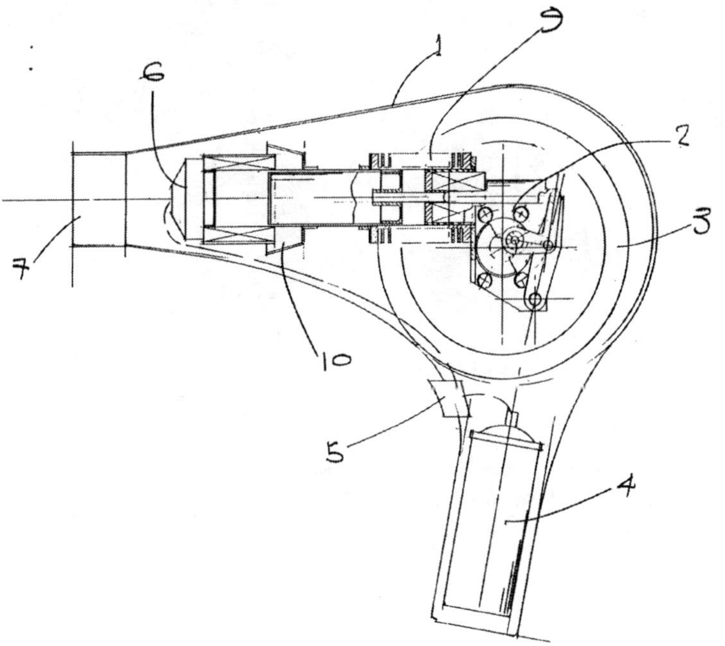

Fig. 7-1: Cordless Stirling engine–powered hair dryer concept drawing. © Allan J. Organ.

I started to work on the engine design in relative isolation from the rest of the dryer, believing that was to be taken care of by Cambridge Consultants. Then I had a meeting with them, and it turns out they had to advise their customer not to go ahead with the idea because of the patent situation. They felt that a Stirling hair dryer wouldn't be patentable because there was a prior patent disclosure. Of course, with a commercial organization, if you can't get exclusivity with something like this, there is very little point in investing in it.

[2] Think of the catalytic converter in your car's exhaust system.

I decided to carry on with the project as an academic exercise because it is so interesting. If one thinks that the performance of a Stirling engine might possibly be extended into this environment, then, it might be the thin end of the wedge for other applications. I like Stirling engines, so that's why I carried on with it. I designed the little engine illustrated in my book *Air Engines.*

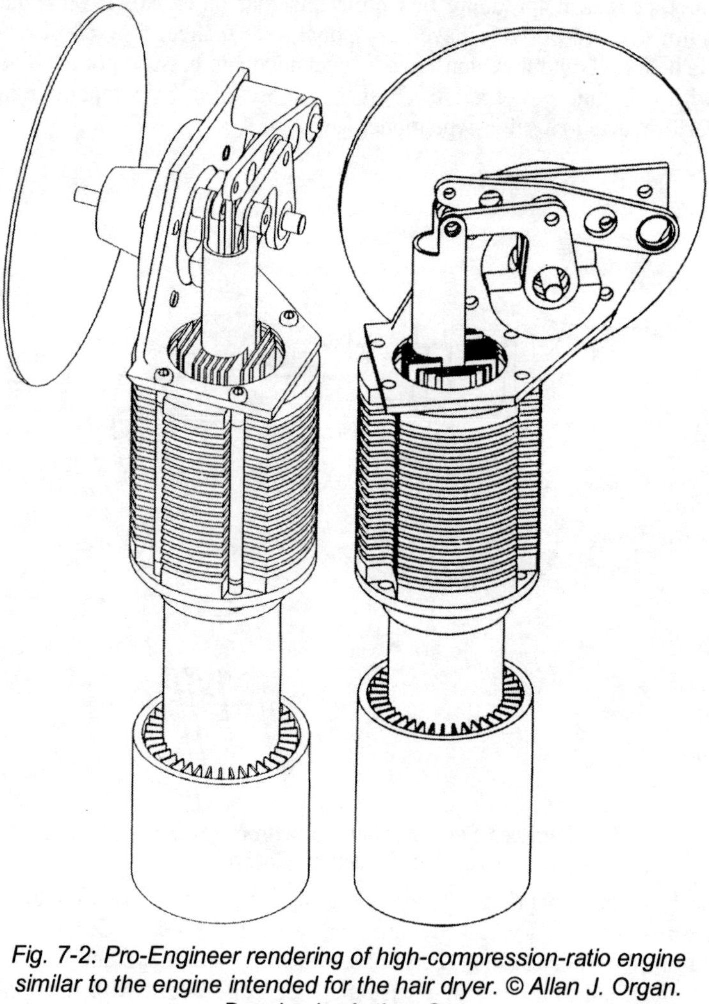

Fig. 7-2: Pro-Engineer rendering of high-compression-ratio engine similar to the engine intended for the hair dryer. © Allan J. Organ. Drawing by Anthea Oon.

We have continued to make changes to the engine since that book was published. We did an engineering study of the whole thing, including the fan, and took it as far as using an aerodynamic optimization program on the fan because in a case like this where performance of the power unit might be marginal, you wouldn't want to build a 50-percent efficient fan if a 75-percent efficient fan was within reach. I programmed Larrabee's brilliant fan and propeller optimization program. This is Eugene Larrabee, a former MIT professor whose work on propellers made a significant contribution to the first human-powered flight across the English Channel.

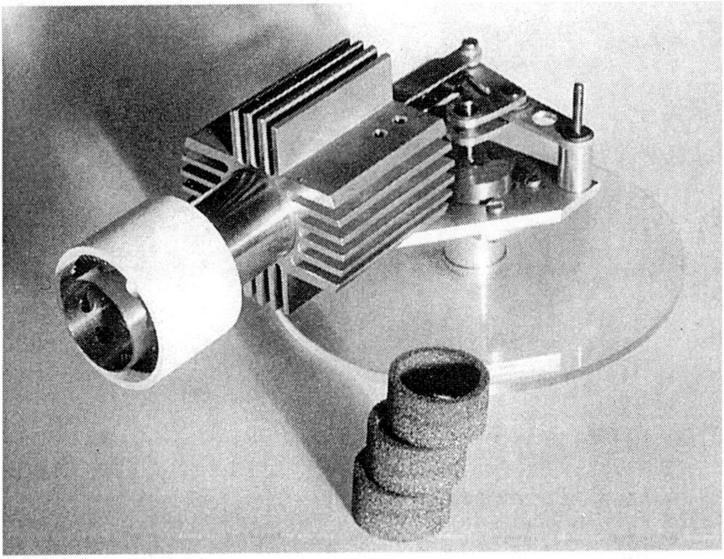

Fig. 7-3: Partially completed hair dryer engine. © Allan J. Organ.

Q: Is the prototype engine running?

Organ: The hair dryer engine itself hasn't run yet. It is in the hands of the person with whom I am collaborating on the mileage marathon engine. The mileage marathon engine, which is a larger version of the hair dryer engine with the same crank mechanism, is running rather promisingly. It powers an ultra-lightweight vehicle which carries a human driver. To make the hair dryer engine run, I think we will have to go for a different form of piston sealing. Oil lubrication of the piston is not compatible with having a blast of air picking up house dust and blowing past the engine unless the engine is sealed in some way. The matter of sealing the engine was never addressed.

The way I think I would go now would be to investigate, theoretically first of all, a pulse engine. As you know, a pulse Stirling engine has just one moving component. It uses acoustic waves inside the engine to shift the gas from the hot end to the cold end and to achieve a Stirling cycle. Pulse engines are the engine equivalent of thermo-acoustic coolers.

I think a pulse Stirling engine would inherently be a better match to a hair dryer because it would have to be a high-speed engine. That raises questions of starting, but I think that starting could be dealt with by using a very simple plastic epicyclical speed increase device and a pull cord.

Fig. 7-4: Stirling-powered Cambridge University mileage marathon vehicle. © Allan J. Organ.

Q: What was the compression ratio that you thought would be right for this particular application?

Organ: The highest we could get, which is why we were using that particular mechanism. I think you want the highest compression ratio you can get. There is some thinking in the literature about there being an optimum compression ratio for Stirling engines. I think that argument is tied to the notion that compression and expansion are adiabatic [without loss or gain of heat], and if you have too high of a compression ratio then you compress up to your expansion temperature and re-expand and don't get any work. I haven't actually followed that analysis with a view to opposing it in any way except to say, that in general terms if your compression and expansion are adiabatic, you haven't got an engine because you haven't got any heat going into it. I don't think one can really use that as an argument against going for high compression. We are using high compression with our mileage marathon vehicle engine, and every time we raise the compression marginally, we're getting a measurable improvement in output.

Fig. 7-5: Lightweight space frame for mileage marathon vehicle. © Allan J. Organ.

Q: It sounds like that approach is worth more investigation.

Organ: Oh, yes. It seems to me that if Stirling engines ever reach the market, they will be in two categories. The first category is the very low compression ratio, high heat exchange area used in highly pressurized engines. These engines follow the Philips and United Stirling traditions. The second category will be high-compression-ratio unpressurized engines, which are extremely simple.

Q: Did you think that this type of engine might be applied to other products such as cordless home heaters. For example, kerosene indoor heaters are readily available but the cordless ones don't have fans, so they don't move the air around the room very well.

Organ: Well, I hope so! I think the whole point about the Stirling engines is that you cannot go into any store in the world and buy a Stirling engine off the shelf. If you could, there would be applications for them. I believe as an academic engineer and therefore not as much of a practitioner, that availability creates demand. How else would we have people applying for a choice between of fifty different sorts of credit cards? We do not need a choice of fifty different credit cards. We just need one. But the marketers come up with some sort of gimmick to market them, and the demand is there.

Fig. 7-6: Stirling engine that powers the mileage marathon vehicle before the latest mods. © Allan J. Organ.

Q: Is thermodynamic efficiency important in an engine used to power a heater?

Organ: No, thermodynamic efficiency isn't much of a criterion for the suitability of the engine in this application. One way or another, you make use of all the heat.

Q: Do you think it would be fair to call the system a 100 percent efficient heating system, because any energy that didn't become noise would become heat?

Organ: It would. But then candles would be 100 percent efficient heaters too. Some way or other, it disposes of all the heat it converts by way of reacting its hydrocarbon fuel with oxygen. From the point of view of designing the engine, I think you have to retain the standard heat engine definition of efficiency, because you want the best possible engine. The engine still needs to be tough enough to drive the fan or you're a bit stuck.

Q: Right, right, you would be stuck if it couldn't drive the fan. Now do you think that this is an engine that could be mass produced economically?

Organ: I think anything can be mass produced once the design is right. I'm not an expert in taking an engine of this sort and saying precisely what I would do to get the costs down while retaining prototype performance and making million of them per week. I have recently examined the insides of a four-stroke, overhead valve engine from the United States, a Ryobi engine for a brush cutter or string trimmer. I would never have dreamed how good the production engineering was in that engine. It's just unbelievably clever stuff. For example, they made up the crank-web from stamped laminations. It's just wonderful. The connecting rod is two stampings, back to back, which hold the cage needle roller bearings in either end for the main bearing. It's all adapted for minimal human intervention on assembly and just knocking them out in presses at one a second or perhaps even at a higher rate.

Q: That's really impressive.

Organ: Yes, indeed. I've had these machines described to me, but I'm ashamed to say I've never taken time to go around a factory and inspect them. One really should. One is not really an engineer until you're aware of these kinds of possibilities.

In the next chapter we will learn about Stirling engines that may be heating and powering homes sooner than you think.

Chapter 8
Cut the Cord to Your Power Company

Interview with Maury White, chief technology officer and founder, Stirling Technology Company, Kennewick, Washington, by the author.

Q: Tell me why you think that using Stirling engines for combined heat and power in homes is a good idea?

White: Our largest initial customer base is in Europe. The interest from Europe is stronger than from North America for three reasons: First, the environmental awareness there is much higher than in North America, energy costs are higher, and most homes are heated with hydronic heating systems, which makes them particularly easy to convert to our residential combined heat and power (CHP) units.

Q: Is hydronic heating what would be called hot water heating in the United States?

White: Yes, in Europe they call them boilers, which are hot water heaters that circulate the heated water to baseboard units in the living space to heat the house.

Fig. 8-1: Free Piston combined heat and power system built by Stirling Technology Company (STC). Marketed in Europe by ENATEC.

Q: I have been told that your method of heating houses and generating electricity is more efficient because you avoid the massive losses between the remote power station and the house. Is that correct?

White: The efficiency is substantially increased by eliminating the transmission line losses. Also when you integrate a Stirling engine with a hydronic heating system, the electricity from the Stirling is generated essentially as a byproduct, so you are getting that electricity without burning the excess fuel that it would take at a central power-generating plant. You're primarily using the gas that would be burned in the boiler system anyway.

It's all integrated, so approximately 98 percent of the heating value of the fuel becomes either hot water or electricity. Our Dutch licensee, ENATEC, has determined that if only 5 percent of the homes in Holland (about 200,000 units) were equipped with the ENATEC CHP system, the net NO_x emissions would be reduced by 100 tons per year, and annual consumption of natural gas would drop by 77 million cubic meters. If these numbers and the corresponding greenhouse gas reductions are extrapolated to all of Europe, where approximately one million boiler units are replaced each year, the environmental benefits are staggering.

Q: So when you compare producing heat and electricity in your house to producing only heat in your house, and producing the electricity for your house at a remote power station, your approach is dramatically more efficient. So do your engines turn on when the house needs electricity, or when the house needs heat?

White: The engines could be configured either way. The ENATEC CHP system is configured to turn on based on heating demand. If there is more electricity generated than is being used by that house, then the power is sold back to the grid through net metering or some similar arrangement.

Q: Where does this type of system makes the most sense?

White: This system will make sense any place where hot water heating is needed for much of the time. The colder the climate, the more benefits the customer gets. But there are also applications of our engines that make sense even in very mild climates. For example, any building such as an apartment building, or an industrial building that uses a significant amount of hot water could make very good use of one of our engines.

Fig. 8-2: Complete STC/ENATEC combined heat and power Stirling engine in enclosure."

Q: Perhaps people in the United States might be interested in your system because it would allow them to keep warm even if a storm took down their power lines.

White: Correct, and if their house is plumbed for hot water (hydronic) space heating it is a very easy retrofit. Of course our system will also heat the hot water for showers, bathroom, and kitchen use.

Q: Why should consumers be interested in your type of heating system versus conventional heating systems?

White: The economics are great! You're getting the electricity without having to consume much net fuel to make it. Depending on your fuel costs, your payback can be as short as three to five years. After the payback period, there is an economic return to the homeowner or in some cases to the utility if they were to purchase the unit and bill the homeowner for the heat and electricity provided. We expect at least a ten-year maintenance-free life, so there are several years of net economic gain. The payback numbers assume the homeowner is replacing his old heating system with one of our heating systems. The economics for new construction can be even better.

Q: In what ways is this technology superior to micro-turbines or fuel cells?

White: Micro-turbines generally come in higher power levels that would be needed for large industrial co-generation applications. I'm not aware of any in the kilowatt range that would be appropriate for residential applications. Fuel cells may someday be a strong competitor, but right now they are several years less mature for this market than our Stirling engines.

Q: What fuels can your engines use?

White: As an external combustion device, it is very fuel-flexible. The most straightforward is natural gas or propane, but we have run engines on direct concentrated solar power and on producer gas from biomass such as wood pellets. Our space engines are heated by radioisotopes, but of course that fuel source wouldn't be used in homes. It is a very fuel-flexible engine, and a burner can be designed for any liquid fuel.

Q: One kilowatt doesn't seem like enough power for an entire house. Should your engines be bigger?

White: The typical power demand for a European home is half or a third of what it is in the U.S., but that is not what determines the size of the engine. The CHP unit is sized to produce the right amount of heat to keep the house warm, and the one kilowatt of electricity is produced as a bonus. This happens to be nearly ideal for the European application. This same engine will produce about 1,250 watts in North America when it is retuned from 50 Hz to 60 Hz for applications where 60 Hz is appropriate.

Q: Is being tied into the grid helpful to the homeowner?

White: We have both grid-independent and grid-intertied units, but the grid connection has a number of advantages for homeowners. Once a grid-intertied unit is installed, the homeowner can forget that it is a Stirling engine. All operation takes place in the background automatically. Stand-alone systems tend to require some attention, although those can be automated as well. In the European application, being grid-intertied really makes sense because to be grid-independent you have to buy a lot of expensive storage batteries, or have to purchase a much more expensive Stirling engine to handle your peak electrical load.

Q: Your Website mentioned the possibility that self-powered appliances such as furnaces can be developed. Where do self-powered appliances make the most sense?

White: One possible example is the pellet stove or even a forced-air furnace. A small Stirling generator integrated in the stove or furnace could provide the power for the fans and the electronics. It could be a completely stand-alone heating appliance.

Q: Would that give you reliable heating even if the electricity went off?

White: Yes. In some parts of the United States, the grid goes down when it's the coldest and you need heat, but even if you have a natural gas or oil furnace, it won't work because there is no electricity to operate the fan and the furnace electronics.

Q: Why do Stirling engines make sense in space applications?

White: They make a lot of sense for deep space probes and planetary missions where solar cells won't work. If you want to send a spacecraft to Europa, or Saturn, or Pluto, it's simply too far from the sun for solar cells to provide enough power to operate the spacecraft. The only solution is to generate electrical power with radioisotope generators. Even some spacecraft traveling closer to the sun, such as missions to Mars or Venus, can't use solar cells very well. In the past, NASA has found that solar panels on Mars had a 30- to 60-day life limit because of dust settling on the solar panels. All previous missions that couldn't make good use of solar cells have used radioisotope thermoelectric generators called RTG's. They are very reliable, but they are low in efficiency.

The Stirling engines we've developed for use in space have about four times the efficiency of the thermoelectrics, so you can get a given job done with one-quarter the amount of radioisotope. This makes launching the spacecraft much safer from an environmental standpoint. The higher efficiency of Stirling engines also enables more missions to be carried out since there is a very limited worldwide supply of suitable radioisotope fuel. Radioisotope fuel is also extremely expensive, so the cost of building deep space probes and planetary landers should go down when our space Stirling engines are used.

Q: Do your systems have a weight advantage over radioisotope generators?

White: Slightly. At the current size level we have a small advantage, but when planned improvements to the Stirling generator are implemented, we will have a significant weight advantage.

Q: You could provide a power system for a Mars lander that would continue to provide power long after solar panels might have become covered in dust?

White: Yes. We anticipate years of useful data collection rather than the one month or so available with solar panels. This also enables mission scientists to explore more interesting regions, such as polar ice caps and deep canyons, which have better prospects for finding signs of life.

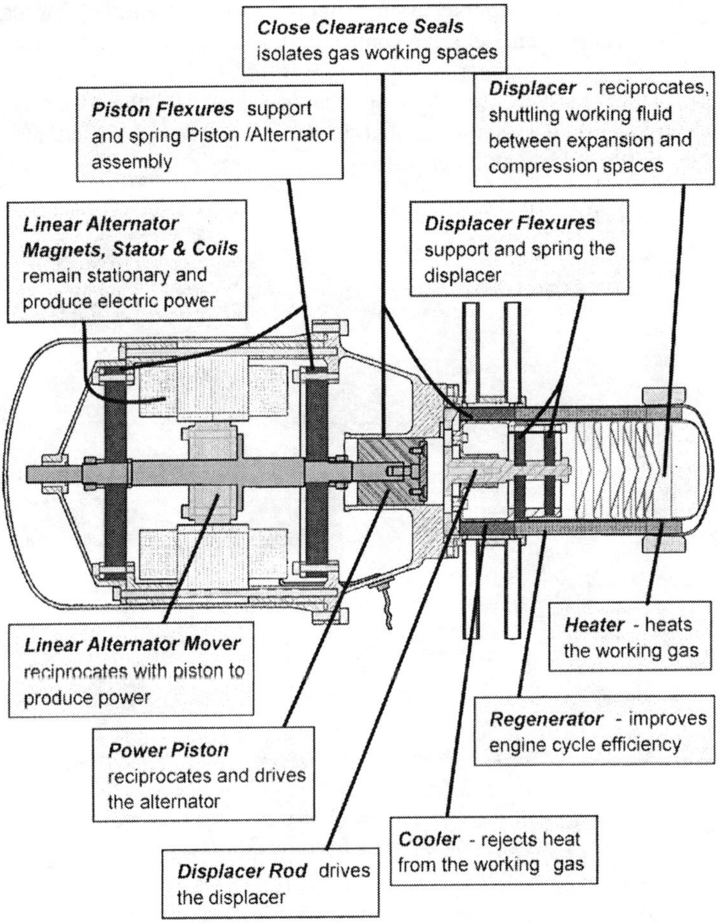

Fig. 8-3: Cross section of Space Stirling engine by STC. See Figure 2-2 for a photograph of a space Stirling engine.

Q: Is this a very low vibration engine?

White: Yes, and the way we package the engines in pairs [see Figure 2-2] makes the entire system have much less vibration than a single engine would. We operate a balanced pair of engines in our space Stirling engines so that the vibrations cancel one another. These are free-piston engines, and they are connected in a back-to-back manner that cancels over 99 percent of the vibrations. The tiny amount of vibration it produces is an order of magnitude below the level that could cause pixel smear for on-board cameras.

Q: Tell me about some of your other products.

White: We have a one-kilowatt gen-set that is really more like a three-kilowatt Honda gen-set. Ours includes a lot of other elements such as a battery storage system, fuel controller, battery charger, cooling system, and engine controller packaged in a stand-alone system. It's fully autonomous; you simply add fuel, and you get both DC and AC power out. Right now it is priced like the expensive prototype product that it is, but it could be mass-produced at a competitive price with other gen-sets.

Fig. 8-4: Limited production generator sets like this one are available at prototype prices from STC today.

Q: So you shouldn't call up and expect to buy it for $500.

White: Correct, although in large quantity production, that's the right general range. That's when we're manufacturing 100,000 a year or more. It's a long way from hand-built individual prototypes. Stirling engines shouldn't be directly compared to internal combustion engines on a kilowatt-for-kilowatt basis for the following reason. A Stirling engine gen-set has about a 5 to 1 advantage over an internal combustion engine/generator, or a solar photovoltaic panel in many applications. If a solar power system is rated at one kilowatt, and you were to look at its average power output over one year, you would notice that it averaged about 200 watts because the rating is at the peak conditions with optimum sunshine levels. When you average in the cloudy periods and night, they generally produce about 20 percent of their peak rating on a twenty-four-hour, seven-day-a-week average, whereas our Stirling engine will put out one kilowatt twenty-four hours a day for years.

RemoteGen™ Product Specifications *

MODEL DESIGNATION	RG-55	RG-450	RG-1000	RG-3000
Electrical Output	60-80 watts; application dependent	450 watts/50 Hz 550 watts/60Hz	1 kW/50 Hz 1.25 kW/60 Hz	3 kW
System Efficiency	29% ***	30% ***	23% ***	39% ***
Duty Cycle	Continuous	Continuous	Continuous	Continuous
Minimum Lifetime	50,000 hrs **	50,000 hrs **	50,000 hrs **	50,000 hrs **
Dimensions	13.8 in. x 5 in. dia	24 in.x 7.5 in. dia	28 in.x 9 in. dia	34 in.x 10 in. dia
Weight	8 lb (3.5 kg)	100 lb (45 kg)	130 lb (59 kg)	160 lb (73 kg)
Maintenance	None	None	None	None

** Specifications are subject to change without notice.*

*** >69,000 maintenance-free, degradation-free hours in ongoing test of similar RG-10 generator.*

**** Burner/heat source efficiency not included—application dependent.*

Q: How would you compare your Stirling engine to internal combustion engines?

White: For IC engines it ranges from 5-1 to 10-1 depending on the application. If a cell phone company needed to build a tower in an area where commercial power was not available, they would typically use a diesel gen-set rated at about ten times the average power requirement. They would then use a big bank of lead acid batteries. They would run the generator a couple of hours a day to charge up the batteries. This approach is used to minimize expensive field maintenance costs, which are required for every few hundred hours of operation. With one of our Stirling engines, the cell phone company could install a 1 kW Stirling engine that covered the average load, and just use a very small bank of batteries to cover peak loads.

Q: How reliable are your engines?

White: Because of our free-piston design and our flexure bearings, coupled with non-contacting clearance seals, our engines are extremely reliable. We have one test engine that has been running continuously for over 8.5 years without any maintenance or degradation! It has only been turned off briefly when the electrical heating element that simulates the radioisotope heat source burned out.

In the next chapter you will learn about a low-tech power-producing Stirling engine that you can build. It will burn anything from corn oil to cow chips.

Chapter 9
A Stirling Engine That Can Run on Corn Oil or Cow Chips

Several times a week, my phone will ring with someone wanting to power their favorite device with a Stirling engine. Their choice of fuel ranges from salad oil to sunshine and from wood chips to cow chips. Usually people want to buy an engine that is priced like a Honda generator, and I hate to disappoint them.

In May 2002 I got a different kind of phone call. The man on the other end said, "Hello! My name is Dr. Arthur Thorpe, and I'm a professor of physics, and I want to know where I can get a Stirling engine that will put out 1 or 2 kilowatts for use in rural Africa. We'd like the engine to be a very economical engine. We want to fuel it with bio-gas that we collect from cow manure. We're having a conference on alternative energy next week, and we would like you to fly to Washington, D.C. at our expense and talk at our conference. If we like what we hear, we'd like to have you build a Stirling engine for us."

Dr. Thorpe is a professor of physics at the university where my father attended medical school. If he had been any other professor from any other university, I would have told him that I was too busy to start any new projects. While I am sure that my father would have been successful in any line of work, the fact that he went to medical school and worked as a highly paid physician in the golden age of medicine made it possible for him to spend lavishly on his children's education. For many years I had felt that I owed my father's university a debt of gratitude.

I agreed to make the trip, and soon I was off to Washington, D.C., where I met Dr. Thorpe, gave my talk on Stirling engines, and worked out the details of the engine that we would build.

Most Stirling engines have problems with their seals. They tend to be pressurized to a very high pressure with either hydrogen or helium. As was mentioned in earlier chapters, hydrogen and helium are very small molecules, and they tend to leak out. The complexity of the engines built by Philips for NASA made me confident that the Philips high-pressure approach was not right for the situation Dr. Thorpe had in mind.

After some consideration I decided to sidestep the problems of pressure and complexity by building a simple engine that we would pressure to a fairly low pressure with ordinary air. It had to be a design that would allow people to run it on the fuel of their choice.

Air is an excellent choice of working fluid for some applications because it is free and readily available. But using air has some disadvantages in a Stirling engine as well. When the hot side of Stirling engine is heated up to flame temperature, the oxygen in the pressurized air inside the engine can cause corrosion, leading to performance degradation and failure of the engine.

One thing that must be avoided always is using pressurized air in a Stirling engine with a splash-lubricated crankcase that has any possibility of getting hot. If you think about it, a pressurized crankcase plus a fine mist of oil plus heat equals an explosion. While Stirling engines are about 99.99 percent safe, you must be careful to avoid a bad design choice when you are doing your initial engine designing concept.

A wet sump with a pressurized crankcase in a Stirling engine can be 100 percent safe if you are using any working fluid that does not contain oxygen. Working fluids such as helium, hydrogen, methane, or nitrogen can be 100-percent safe engine with or without a wet sump. Using pressurized air as a working fluid can be 100-percent safe if there is no oil mist in engine that can get hot. Wet sumps are not a good idea in pressurized air engines.

If I had been planning on powering this engine with natural gas, I would have considered using methane (natural gas) as the working fluid in addition to the fuel. Methane has some significant thermodynamic advantages as a working fluid over air.[3]

Using hydrogen, can be very safe inside a Stirling engine. The reason is that when the hydrogen is inside the engine, there is no oxygen inside the engine and hence no possibility of an explosion. The worst that could happen is a leak of the extremely small amount of hydrogen into the atmosphere.

Because we wanted a low-cost, simple engine, I decided to use air as the working fluid in our engine. I returned home from Washington, DC and began rereading my Stirling engine books for design ideas.

In any mechanical engineering project, it is extremely unlikely that you will build something that is 100 percent new. Someone somewhere along the line has probably built something very similar to what you have in mind.

As I began to reread my Stirling engine books and make notes in my engineering notebook, I came across the name of L. Merrick Lockwood, who in the early 1980s worked with the Asia Foundation to develop a Stirling engine for use in rural Bangladesh. The engine that Dr. Lockwood had developed (funded by the US Agency for International Development) was intended to be simple to

---

[3] *The Stirling Alternative*, by G. Walker (Editor), Graham Reader, Owen R. Fauvel, Edward R. Bingham (Taylor & Francis, October 1994).

manufacture, reliable, and economical. The intended fuel was rice husks.

I decided to try to contact Dr. Lockwood. I sent an email to the last address I could find for him, but it bounced. Disappointed, I decided to call my friends in the Stirling engine community to see if anyone had a more current address for L. Merrick Lockwood. Fortunately, someone did have an address for him. He was now living in India, but they didn't have a phone number or an email address. So I sat down and wrote him a letter. Writing a letter to someone in India felt like going back through a time warp to an era, when paper, pens, and stamps were the main methods of world wide communication.

I put a stamp on my letter, dropped it in the mail, and thought, *That was probably a waste of time.* Ten days later I sat down at my computer, logged in to my email account, and saw an email message from L. Merrick Lockwood in India. I was delighted!

After a series of phone calls and emails, Dr. Lockwood agreed to help us with our project because the goals for our project were substantially similar to the goals for the project that he had started in the 1980s.

The engine that Dr. Lockwood had helped build was a low-tech, simple, slow-turning engine. It had a relatively large displacement, and the design was simple enough so that all the parts could be fabricated in Bangladesh. The engine is a one cylinder design of *beta* configuration (the displacer is in the same cylinder as the power piston). The fuel for the engine he developed was rice husks.

If you aren't familiar with rice farming, rice husk is the outer portion of the grain that surrounds the part that you eat. This husk has to be removed from the rice before it is good to eat. In areas that grow a lot of rice, getting rid of the rice husks is a problem. Dr. Lockwood's concept was to develop a burner that would burn rice husks to power the Stirling engine. This engine was intended to help the entrepreneurial activities of the people in Bangladesh.

It was an excellent concept, and several prototypes were built and tested. Unfortunately Dr. Lockwood never was able to complete the design to the level of refinement that he had in mind due to a lack of funding. Fortunately the design of the engine's hot section was general enough so that a burner could be built to burn other fuels easily.

Fig. 9-1: A low-tech 5 hp (3.7 kW) rice husk–fueled Stirling engine.

Fig. 9-2: Worker applies insulation to heater of the rice husk engine in Bangladesh.

The first prototype of the rice husk–burning engine was built by Sunpower, Inc. in Athens, Ohio. It used a drive mechanism called an Ericsson linkage. This complex mechanical arrangement, derived from the Ericsson hot air engines in the early 1900s, turned out to be the weak point in the design.

When the engine was operating under load, everything was just fine. But when the engine was allowed to rev up to its free-speed capacity (which would happen when it was not under load) the entire power output of the engine was being used up to move the displacer back and forth. This caused tremendous loads in the displacer linkage, and it would fail. With the help of Dr. Lockwood, we began a detailed design of our Stirling engine project. It would not use a complex mechanism; it would use a simple double-throw crank.

Dr. Lockwood's original engine had operated with dry pistons sealed by Teflon and graphite-based O-rings. Since this worked well on his engine we decided to use it on ours.

Dr. Lockwood was a tremendous help in this project. We learned a lot from him about what would work and what would not. He even told us about some amazing metal-forming methods he had tried in Bangladesh. For example, the hot cap of the engine needs to be a dome shape to reduce the stress in the material. Typically, most shops in the United States would fabricate this by metal spinning. Dr. Lockwood tried spinning the metal on a lathe, but it turned out that metal spinning would harden the stainless steel. This required several additional annealing steps to soften the stainless so that it could be processed to the next stage.

Dr. Lockwood developed metal forming methods that we suggest that you *not* try at home. To build the dome in the hot heater head, he built cast-iron dies. He then explosively formed the hot cap using locally produced fireworks! We definitely do not recommend this approach, but it is worth knowing that it worked and was a very effective method for forming the hot head of the engine. Due to Dr. Lockwood's careful safety precautions, no one was ever injured. Our engine is about a foot in diameter (30 cm), about four feet (1.2 m) long, and is expected to weigh seventy-five pounds. If this seems large for an engine that puts out a small amount of power, remember that this engine is effectively a refinery plus an engine. An internal combustion engine is fairly useless if you try to run it on wood. This Stirling engine can run on anything that gets hot and has its refinery built in. You should be able to burn biomass, propane, or any other fuel for which you care to design a burner. As I write this chapter, our Stirling engine is under construction and is expected to be delivered to the university shortly after this book goes to press.

Fig. 9-3: Explosive metal forming of Stirling engine parts in Bangladesh. Note the water column. Don't try this at home!

Fig. 9-4: Dies used in explosive metal forming

Fig. 9-5: The finished metal dome ready for the next stage of assembly.

Will this engine be simple enough to build in a home workshop? Probably so. It certainly was simple enough to be built in a relatively low tech situation such as Bangladesh. Of course not every home machine shop will have the appropriate tools, but most small machine shops in the United States could fabricate this engine. If the thought of overhauling a V-8 engine doesn't scare you, you probably could build this Stirling engine. If you were willing to take the time to learn the necessary skills, it might be a lot of fun.

American Stirling Company will be publishing a book by Dr. L. Merrick Lockwood on this engine, which will include detailed photos, drawings, and plans. That book will not only include plans for the original engine but also will include plans for our updated, somewhat simplified version. Yes, you probably can build this engine if you want to! Please check our Web site at stirlingengine.com for ordering information and additional details.

In the next chapter you will learn how a mountain man built a 2.5 hp wood-fueled Stirling engine that he uses to heat and power his shop.

Chapter 10
The Mountain Man's 2.5 hp (1.8 kW) Stirling Engine

An interview with Jim Symanski, Garden Valley, California, by the author.

Q: How did you get interested in Stirling engines?
Symanski: After I got out of high school, I met a friend of mine who was as interested in metal work as I was. I was over at his place one day, and he showed me a little tiny Stirling engine built by Solar Engines in Phoenix, Arizona.

Well, I got interested in them, and I figured I might try to build some of these things. I had a very hard time getting the first one I built to run, but I finally got it to run. It wasn't very good, but it worked. About a year or so later, I made a little fan Stirling engine. That one worked pretty well!

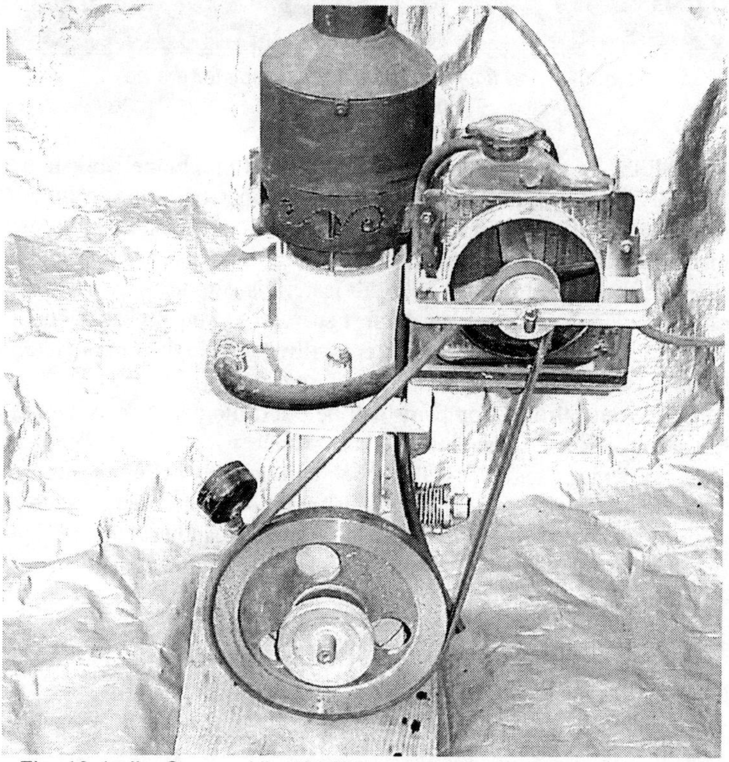

Fig. 10-1: Jim Symanski's 100 Watt continious 400 Watt max power Stirling engine.

A couple of years later, I built one that had been featured on a TV show. It was a copy of a four-inch Ryder. There was picture of it in the book on Stirling engines that my friend had. So I looked at the picture, and I just built it from the picture. And it worked fine! It worked great! In fact, it's still running. I run it all the time.

Q: Did you measure the power output of that engine?

Symanski: It puts out about $^{1}/_{5}$ of a horsepower at 350 rpm. It probably runs about three times faster than the original.

Q: How did you fuel that engine?

Symanski: I had a propane burner on it with a fairly big O-ring on it. It's pretty big. I made a lot of modifications to it until I had it running the best that I could. I was pretty happy with it. Then I figured I'd try to build a pressurized engine, so I did some experiments to figure out how much heating surface and cooling surface you need, and I got that one running but it wasn't pressurized. Then I built a pressurized one. The little one I have now puts out about 400 watts at about 600 rpm.

Q: What is the working fluid in that engine?

Symanski: Air.

It was a *gamma* configuration. I fabricated the crankcase out of a piece of steel pipe, and it held the pressure that I wanted.

By that time, I had much better machines and a better lathe, and I machined it all. It has a wet sump in it, so it lubricated itself. I put the air compressor on it so it pumped itself up. I just added a simple stainless steel hot cap to it, with no grooves or fins, but the cooler side was a problem. I kept experimenting with it until I finally got 400 Watts (.54 hp) out of it.

Q: That's pretty good!

Symanski: Yeah, it's not too bad! With about forty-five pounds of pressure it was running pretty well. About six months later, I decided that I ought to build a little bigger one to run my shop, so I started working on a bigger engine. I used what I already knew about the heater and the cooler and just made it really big.

I saw a model of an old Stirling that Robert Stirling had built before he built his 45-horsepower engine. I just modeled my engine after that one. It was simple. It didn't need a crankcase because it had to a double acting power piston. It has two displacers. I worked on it for about six months or so, and I finally got it right. I got it running, and I found out I had to put a governor on it or it would run away. So I put a governor like many steam engines have, on my engine. I did a lot of testing on it, and I got it up to a maximum of two-and-a-half horsepower.

Fig. 10-2: Jim Symanski's 2.5 hp (1.8 kW) wood-burning Stirling engine.

Q: What was the working fluid?

Symanski: At first it was just pressurized air. Then I started experimenting with helium, and I put everything in there. I put in CO_2. I put in argon. What I found was that the helium gave you maybe a little bit more horsepower but not much. The engine ran with a little more snap, and it was more lively with helium than with air. So helium was the best, and the air was the next best. The CO_2 was a little worse than the air, and the argon was probably just as bad as the CO_2 was. I'm using air in there now, and it runs fine.

Q: What is the engine's basic configuration?

Symanski: Right. It's got two vertical displacers with the hot caps side-by-side inside the firebox. Then there's a beam on top that operates it. It displaces up and down because they are 180 degrees apart. One goes up while the other one is coming down.

Q: So it does not have a pressurized crankcase.

Symanski: Right, in fact it has no crankcase at all. The power piston is powered right off the top between the two displacers.

Q: How does your speed regulation mechanism work?

Symanski: I have a valve between the two displacers. The valve is operated by a twin ball governor mechanism like you have probably seen on the old steam engines. All the valve does is relieve air pressure from one side of the piston to the other.

Fig. 10-3: Speed regulation governor.

Q: It allows the gas to bypass the power mechanism so you can achieve the speed that you want.

Symanski: Exactly! Since the engine is wood fired, it's almost impossible to regulate the fire quickly enough for any speed regulation. The hot caps are mounted vertically in the top of the firebox.

At first I wasn't getting enough draft so added on a forced air system for the firebox, so now I get a hotter fire.

Q: Is the forced air system powered by the engine?

Symanski: Yes.

Q: Sealing Stirling engines has sometimes been a difficulty. How do you seal the engine?

Symanski: I use ordinary buna rubber O-rings, just the regular multipurpose ones that everybody uses. The displacers use Quad O-rings. Some people call them the "square O-ring." It's almost like two

O-rings. Those things work great! I've never touched them. I've had helium in that engine for months. I've not had any leaks from those O-rings.

Fig. 10-4: Firebox for 2.5 hp engine.

There was one leak where the power piston was with the single O-ring. Now I've talked to people, and they say that buna does not seal helium very well. Well, I've had it in there under 200 pounds of pressure, and it very slowly does leak down. But it holds quite well!

Q: What is the air pressure that you are using when the engine is putting out 2 $^{1}/_{2}$ horsepower on air?

Symanski: The peak pressure is about 250 pounds.

Q: How do you pressurize the engine?

Symanski: I have an air compressor on the engine. The engine will start with atmospheric pressure, but it has to have about 50 pounds in it to get enough power to run the air compressor to pump the engine up to operating pressure. When the pressure starts getting up to about 100 pounds, then the engine really starts generating power. The air compressor pumps up to about 130 pounds. You actually get more than that in there, especially when the engine is cold pumping itself up. What happens is that the engine still can accept air at the low-pressure point of its cycle.

Q: How do you get the full 250 psi inside the engine?

Symanski: The 250 pressure is the peak pressure inside the engine. I pressurize it to 130 psi with the compressor, and the temperature changes inside the engine produce a peak pressure of 250 psi.

Q: How is the engine cooled?

Symanski: It's water cooled. I have a radiator like they used to use on one of those old-time engines. It drips the cooling water over the top of a screen. As the water flows down the screen the air cools it. You get some evaporative cooling too. It's very effective.

Q: How do you use the engine in your workshop?

Symanski: I'm using it to run my shop right now because I don't have any electricity at my current location. I've got a 40-foot line shaft running the length of my shop. I have a clutch in the line shaft so I can engage it and disengage it. I connect up whatever machine I want with a flat belt.

Q: Isn't that how things were done in factories before they switched to having an electric motor on every machine?

Symanski: Exactly! This is a lot more efficient way of powering a shop if you only have a small amount of power. Because you're using mechanical power you don't have the conversion losses between generators and motors, and the machines run well.

Q: So is $2^1/_2$ horsepower enough to run your lathe and your milling machine?

Symanski: Right, it runs them with no problems.

Q: There are several different ways to measure power output of an engine. How did you do it?

Symanski: I used a simple strap brake on the flywheel. I used a strap like you would use to tie a load on a truck down, and then screwed a bunch of wooden blocks to it so the wood blocks became the contact surface with the fly wheel. I attached a spring scale to the floor, ran the strap with the wood blocks over the flywheel and on the other side I kept adding weights to slow your engine down. At the same time I was accurately measuring the engines rpm's with a tachometer.

I just kept loading weight on until I brought it down to where I got the maximum pulling power out of it. Then I measured the difference between the weight the engine was lifting on one side and the spring scale to figure out the torque. Once you have the engine's torque and speed, you can easily find the horsepower. It's dead-on accurate; you can't get any better than that.

Q: That's certainly a very good method for measuring output power.

Symanski: I know it's a good method. The only problem is that when you start getting into higher horsepower outputs, it's difficult to keep the brake working. When it starts heating up you'll have problems with it. It can start slipping or burning too. When it gets too hot you have to adjust it all the time.

Fig. 10-5: Heater detail on 2.5 hp engine.

Q: What do the heaters and the coolers on the 2 $^{1}/_{2}$ horsepower engine look like?

Symanski: On the outside of the hot cap, I have fins cut lengthwise into a plain steel cap. They are quite large; they're probably about $^{3}/_{16}$" [4.7 mm] wide. I don't want to make them too small on the outside because of the wood fire.

Q: Would they burn up in the wood fire if they were finer fins?

Symanski: Well, not only would they burn up, but if you get any kind of carbon or anything in there, they clog up. Small fins on the outside are not a good idea if you are burning wood. Your power drops off right away if you get any carbon coating on the hot cap. You have to keep the fire at a high heat to keep the carbon burned off.

You have to keep a very clean running fire for the engine to keep putting out power. If the power starts dropping off, you know right away that you have some soft wood in or something and there is carbon on the hot caps. If you heat the engine back up and burn that carbon off, the power will come right back again.

Q: Do you remember what alloy of steel the hot cap was made from?

Symanski: I have no idea. I think it started out as an old piece of steel pipe with a $^3/_4$" wall that was extra, extra heavy. On the inside of the hot cap there are machined fins running lengthwise. I think they're about $^1/_{32}$" [0.8 mm] wide and about $^5/_{16}$" [8 mm] deep. The grooves are about 8 or 9 inches long (20 to 23 cm), and I think there are about 110 of them around the inside of the hot cap.

Q: What is the cooler like?

Symanski: The cooler is made out of aluminum, and it's bigger in diameter with a lot more fins both internally exposed to the hot air and externally in the water jacket. It's much bigger in diameter than the hot head. The ratio of cooling surface to heating surface area is something like one and three-fourths times more cooling surface than heating surface. I figure that's not really enough; there should be more cooling surface. But for this engine, it works OK.

Q: If you were to build another engine, what ratio of cooler surface area to heater surface area would you use?

Symanski: In that style engine, I'd probably go up to 2 or a 2 $^1/_2$ ratio, and I wouldn't use aluminum again. I would use copper.

Q: Copper is a much better conductor than aluminum.

Symanski: You're right. I was looking for some copper when I built this engine, but I couldn't find any copper pipe with a thick enough wall, so I used aluminum, but copper would have been better.

Q: Have you drawn up plans for this engine?

Symanski: No, I never do. When I build something I just build it out of my head. Everything I build is cut and fit. I'm usually limited to the materials I have around or materials I can scrounge up. So I didn't spend a million dollars for materials; I just put it together.

Q: It's very impressive that you can make enough power to run your own machine shop.

Symanski: The only limiting thing is stoking the fire all the time. I have to stoke the fire every 15 minutes. I have a small firebox. I didn't want a big firebox in it because the smaller the fire you've got, the hotter you can run it.

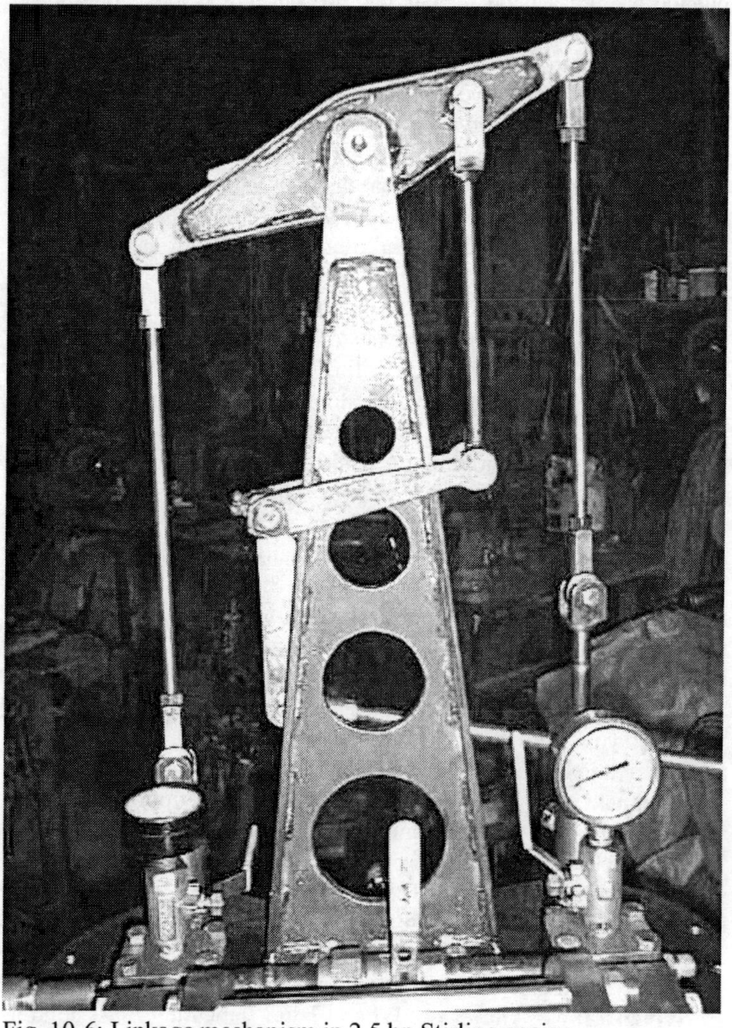

Fig. 10-6: Linkage mechanism in 2.5 hp Stirling engine.

Q: What would you do differently if you were to build another engine for the same purpose again?

Symanski: Oh, I'd build a bigger heating and cooling surface on it and then it would run at a little bit more pressure. If I could get a bigger heater and cooler, then I could pull more rpm, and that would help a lot.

Q: When the engine is putting out 2.5 hp (1.8 kW) how fast is it running?

Symanski: It's running about 213 rpm when we measured the maximum power output.

In the next chapter you will learn how an Oklahoma auto mechanic has built a Stirling engine that powers the rotisserie in his barbecue grill. He calls it the "Chicken Flinger."

Chapter 11
A Stirling Powered Rotisserie
the Chicken Flinger

An interview with Craig Roop, Tulsa, Oklahoma, by the author.

Q: How did you get interested in Stirling engines?

Roop: The very first Stirling engine I saw was a Lake Breeze fan that an older gentleman who owned a motorcycle shop that I frequented had in his office. He would run it every time we came in. Then I saw an ad in a Home Shop Machinist magazine from Bob Bailey at Bailey Craftsman Supply for a Stirling engine kit, so I ordered the kit.

It took ten minutes or less to put the engine together, and it fired right up and ran. I thought, *Well, I can make this,* and so I went and drew all of my money out of my savings account and started buying machine tools that I had no idea how to run.

Q: So you had never run a machine tool before you built your first Stirling engine?

Roop: Well, I've been a mechanic all my life, but I had never run any machine tools other than a brake lathe that mechanics use for repairing automobile disk brakes, and a drill press.

Q: What did you buy first?

Roop: The first thing I bought was a 9" x 20" [23 x 51 cm] Taiwan-made lathe and less than a week later, I bought the equivalent Taiwan-made mill drill. The week after that, I bought a 7" x 10" [18 x 25 cm] mini-lathe. And then I bought two six-inch (15 cm) grinders and a drill press, and I started buying tooling and making tooling. A friend of mine who is a commercial machinist had lots of good ideas about how to make things. He was always saying, "Here's how you can do that real cheap!"

Q: So tell me about the first Stirling engine that you built on your own.

Roop: The first engine I built on my own was a low temperature–difference engine. I built it by looking at pictures on the Internet and reading about stuff on the Internet. It was made from an embarrassing array of recycled materials, but it still runs today, so I can't complain.

Q: What kind of parts did you use?

Roop: The aluminum plates came from a traffic sign that got knocked down in a car wreck. And the displacer chamber plastic was part of a gravy separator. The power piston cylinder is a piece of brass tubing from a 1989 XJ-6 Jaguar heater core inlet tube.

The piston was made from a piece of electro-discharge machining graphite material that a guy at a machine shop gave me. I actually made that piston in the drill press.

Q: That's very impressive. How did you manage making a piston in a drill press?

Roop: First I drilled a hole through my piece of graphite. Next I fastened a bolt through that hole and tightened the bolt in the drill press. Then I clamped a vise to the drill press table and put a screwdriver in the vise, just kind of loosely. Then I turned the drill press on and simply tilted the screwdriver up and down, and manipulated the thing just until I got the piston made. That just kind of roughed it out, and I finished the piston by hand with just an emery board. [This method is dangerous and is not recommended -- ed.]

Q: That's very pretty impressive!

Roop: Yeah, well, I was desperate.

Q: Did your engine run the first time?

Roop: No, everything was way too tight. So I kept going back and kept modifying the engine and messing with stuff. I think I had to make two pistons before I got one that actually worked. Then it worked pretty well!

Q: What did you build next?

Roop: I built a couple more small, low temperature–difference engines with a 1 $^{1}/_{2}$" (3.81 cm) diameter displacer. These were fine, but they really didn't trip my trigger. I liked them, but I wanted something that ran faster and made more noise or something.

Next I built a little beam engine, and it ran very well and really fast too. Then one day I was throwing away some junk automobile parts where I work as an auto mechanic and I found this shock. Inside the shock there was a very smooth tube that I used to build the Chicken Flinger engine. I took the tube directly to a welding shop and had them weld a cap on it because I knew that I couldn't weld it without destroying it. The bore of this tube was perfect—it was so nice and it was already polished. I knew that I would never be able to machine something like that.

Q: Tell me how you came up with the idea of building a Stirling engine to power the rotisserie in a barbecue?

Roop: Well, I was cooking in my backyard with a propane grill that had an electrically powered rotisserie, and I tripped over the cord and just about broke my neck. That's what gave me the idea because I knew that the power requirements for a rotisserie were pretty small because it used a gear reduction.

I already had built a small engine that had the displacer and the power piston in the same cylinder, and up until then I had been content to have it sitting here on the table running on a little bitty alcohol burner. But right then I knew what I wanted to use it for.

So after tripping over the barbecue cord, I took the rotisserie gearbox apart, cut a little opening in the side of the gearbox, and took the electrical part out completely intact. Next I mounted the Chicken Flinger in the barbecue grill so that the flame would come in contact with the hot head of the engine, and I drove the original gearbox with a small O-ring belt that I made.

Q: OK. Tell me a little bit about the type of grill you put this in. Is it hooded or is it open?

Roop: It is a hooded Model 100 Sunbeam propane grill. It's probably one of their least expensive units, and I've had it forever.

Q: So it's a Sunbeam grill that had an electric rotisserie in it. Did the motor you took out say how powerful it was?

Roop: It was a .25 watt electric motor. You really don't need much power to run the rotisserie. Everything I read about model Stirling engines said, they don't make much power, but to power this rotisserie you don't need much power. I ended up using a larger pulley on the engine and doing some burner adjustments so that the engine wouldn't over-speed the gearbox.

Q: Tell me about the process of retrofitting the barbecue.

Roop: There were just four screws to take the electrical gearbox apart. As soon as I got it apart I lifted the electrical part of it right out completely intact—cord and all. Then I took a pair of tin snips and cut a little window about $^3/_4$" [19 mm] wide and about 1" [25 mm] or so long on the correct side of the gearbox. Before I put it back together, I put in my O-ring belt to connect the engine to the gearbox.

The hot cap sticks up right next to the burner flame from the bottom. The crankshaft sticks out to the left, to the side, so it can line up with the rotisserie gearbox. The engine is running vertically.

Q: So did you use the existing burner or did you modify it in some way?

Roop: I actually took a piece of $^3/_{16}$" [4.8 mm] copper tubing and bent it in a U shape to curve around to within about $^3/_8$" [9.5 mm]. Then I cut a couple of little tiny slots in it and then drilled the burner itself to fit the copper tubing and then mounted it in there so I now have a ring burner around the hot cap of the engine.

It would be nice to have a separate burner so you could control the speed of the rotisserie separately from your cooking flame, but this works well.

Fig. 11-1: Close-up of Chicken Flinger Stirling engine installed in grill.

Q: Tell me about the first time you ran this engine in the barbecue.

Roop: It was the fourth of July of 1999, and I decided I wanted to cook a chicken. The girl I was dating at the time hadn't seen this little contraption, and when she saw the engine running, she said, "You're going to fling the chicken off. You're making a chicken flinger." So she was the one who named it "The Chicken Flinger."

After that when we would eat a barbecued chicken together she would tease, "How's your Flung Chicken?" It was always delicious.

Q: Did it work well the first time?

Roop: Oh yeah, it worked great! Alice had suggested that I put a couple of pieces of potato and small piece of apple in the chicken to help make sure it stayed moist inside. That's a very good idea. When the potatoes are cooked, you know the chicken is done.

The engine works great. The day before Thanksgiving, 2000, I took the rotisserie to work and cooked chickens for all the guys at the shop. They thoroughly enjoyed it because everybody and their family was there too. The kids just loved it! You know this one little boy sat there and just watched it. "It's speeding up! It's slowing down! It's speeding up!" It was so funny!

Q: Stirling engines have to be hot on one side and cold on the other side. How you keep the cold side of the Chicken Flinger engine cold?

Roop: It uses disks from old computer hard drives that I bead blasted and bored to be a real snug fit on the cylinder. I made little spacers to hold them apart. Believe it or not, it's all JB welded [a 500° F or 260° C epoxy] into position in the right place.

Q: Has the engine continued to run well since the first time you used it?

Roop: Yes. I replaced the displacer shaft bushings one time, probably within a year after I built it. They were all used. I don't know if they were more used than the other ones, but I haven't had any more problems with it since then.

Q: Tell me some more stories that you have about what you've cooked and the people you've cooked for.

Roop: I rotisseried a turkey this year for my ex-father-in-law and his mother. Most people don't listen when I've told them that my Stirling engine powers the rotisserie on my barbecue grill, but when they hear this noise and they realize that the Stirling engine actually *is* running the thing, then they are really impressed. **Q:** Have you done other meats besides chicken and turkey?

Roop: Oh, yes. A roast turned out really well. Of course when I'm using the rotisserie I'm usually either doing chickens or turkeys. I did try some shish-ka-bobs once, but they came apart. There was my dinner down in the bottom of the grill.

Q: Even though it's a propane grill, do you use any hickory or mesquite wood in the bottom to give it a better flavor?

Roop: It depends on what I am cooking and how I season it. Usually because of the way things are in a rotisserie, you need a pan underneath the chicken. I used a piece of stainless steel and made a deflector so that any drippings run off to the side of thing and not onto the propane flames so that you don't get a great big fire. That way it never scorches the meat. You might have a little bitty fire from the chicken dripping but that fire will burn off and it's deflected away from the meat.

Q: Do you have a favorite type of barbecue sauce?

Roop: I use KC Masterpiece brand of barbecue sauce. It's great.

Fig. 11-2: Grill with hood closed showing location of Chicken Flinger engine.

Q: How long does it take you to cook a turkey or a chicken?

Roop: An average sized chicken, about two to three pounds [0.9 to 1.3 kg], takes about two hours. A seven to nine pound (3.2 to 4.1 kg) turkey takes about four to five hours. I have a tendency to cook them pretty slowly. You can do a roast in an hour and fifteen minutes because you want to turn the heat up a little bit more on a roast than you would for a chicken.

I always use a meat thermometer on a chicken. That's another good thing about using the potatoes inside the chicken: if you check the potatoes and they are still raw, the chicken is not cooked. The potatoes need to be cut up into 1" [2.54 cm] cubes.

Q: It seems like your engine will run as long as you need it to. How many hours have you put on it since you built it?

Roop: Believe it or not, I used to keep track. I know that there's well over a hundred hours of run time on this engine. I've cooked a lot of things with the Chicken Flinger. One time I did an endurance test over a weekend and ran the engine for almost twenty-four hours straight. I turned it off whenever I went to bed. I ran it on an alcohol burner. I used two quarts of alcohol that one weekend just running

Fig. 11-3: Unassembled Chicken Flinger Stirling engine with 6" (15.24 cm) ruler shown for scale.

that one Stirling engine. The longest it ever ran without stopping was about seventeen hours.

Q: So it's a pretty well developed engine for this application and it seems like it's going to last.

Roop: Yes, I think so.

Q: How much did it cost to build?

Roop: Probably about twelve dollars.

Q: You must be very good at finding surplus parts.

Roop: Yes. It cost ten dollars to have the cap welded on. The other two dollars was for some screws that I needed. Everything else came from old computer hard drives. The ball bearings came from computer hard drives. The power piston insert is part of a Volvo brake master-cylinder. The displacer is a cigar tube, and all the shafting came from old computer stuff. Of course you could go out and buy all these things, but I'm just a scrounger.

Q: That's a good approach.

Roop: Sure, and anybody can do it. Because this engine seemed to work so well and I enjoyed building it so much, it just hit me that part of the fun of building this engine was building it from recycled materials. After that I made a real effort to try to build everything from recycled materials as long as it didn't hamper performance. If I needed new ball bearings, I'd break down and go spend the four dollars and get new ball bearings. But if I didn't, then I wouldn't.

Q: You found computer hard drives to be a pretty good source of ball bearings?

Roop: Oh, yes. Old VCRs have wonderful ball bearings. Old CD-ROM drives have some really nice ones. All this old electronic stuff that has very high quality ball bearings. I made a very simple little tool that is shown in the plans that anyone can make to remove ball bearings. It's just a tiny two-jaw puller.

Q: Could this engine be engineered so that it would be extremely robust so that people didn't have to take any special care of it?

Roop: Yes! In fact I have the prototype finished for that particular endeavor. It has a completely enclosed crankcase. It ought to be a little stronger, and it definitely will be a little quieter because I got rid of a ball bearings in one place and put a bronze bushing in.

Q: That's pretty good! Have you ever directly measured the power output of this engine?

Roop: Not accurately. I have hung a Starrett brand direct-reading tachometer on it, and when I'm running the engine on a propane torch it turns about 2100 rpm with no load.

Q: Is there anything else about this engine that our readers would be interested in?

Roop: The engine has two separate crankshafts with a counter-rotating gear drive that helps balance the engine. You are able to independently balance the two reciprocating motions, even though the pistons have different weights. You can also easily make timing adjustments for optimal performance. While I don't have any scientific data to back it up, this engine doesn't vibrate across the table as much as some other model Stirling engines.

Also if anybody is going to try using recycled shock absorbers, they need to be really careful about doing it. Because when that one blew up on me, it really scared me. When it blew apart it knocked an 8-foot [2.5 m] long double fluorescent light fixture right off the ceiling and then dented the side of the steel industrial building I was working in.

I was using an industrial cutoff tool, and that set it off. I had already vented the pressure on one side, but I didn't realize it had applied pressure on the other side of the shock piston too. The sparks from the cutoff tool ignited the gas or the vapor inside the shock, and it went off like a cannon.

Fig. 11-4: Complete Chicken Flinger engine is shown above. The upper engine is a new version with an enclosed crankcase. The new engine does not yet have its cooling fins installed.

Q: Is there a safe way to salvage an old shock absorber?

Roop: Now I take the shock outside because the oil is kind of nasty. I use the edge of a flat file, and with the area that I am filing on pointed away from me I start filing on it until the file just breaks through the surface and the oil pressure bleeds off. After I've bled the pressure off of there, I'll go to the other end of the shock and do the same thing. If you can move the shock piston back and forth really easily, then you know there's no pressure in it. After the pressure is gone I use a hacksaw and make a very slow cut around the circumference of the shock. After the bottom of the shock is cut off, it will disassemble itself. Inside the shock is a little tube that you are after.

Q: Do most shocks have the type of tube someone would need for a project like this inside or do people need to find a Jaguar shock?

Roop: Most shocks have an inner tube. You don't need to find a Jaguar shock, but you do need to find one with an inner tube. Almost all McPherson strut shocks have an inner tube; however, most American car shocks, like Monroes and ones like that, are single tube designs. European car shocks are often double tube shocks.

You want to find an inner shock tube that is about one inch (2.54 cm) in diameter

You may have to go through several shocks to find one that will work, but McPherson strut assemblies are almost always double tubes, and they are a good place to start.

Q: What other advice do you have for someone who wanted to build an engine like this?

Roop: I like lubricating my engines with Marvel Mystery Oil [a brand of lubricating oil sold for general use]. If you want to run your engine totally dry, I would make a graphite piston which would then be self-lubricating. If you want to use a steel piston you will need a little bit of lubrication. I've tried Mobil 1 brand and all the synthetics, and I've gone back to Marvel Mystery Oil. It's super thin. It penetrates well, and it doesn't smell bad, which is important for an engine that's going to be around something you are going to cook.

I probably put a couple of drops of oil on the Chicken Flinger whenever I start it up, and after I'm done cooking something, I'll put another couple of oil drops on it before I put it away.

Fig. 11-5: Almost ready to serve Stirling "rotisseried" chicken. Note the O-ring belt going down to the Chicken Flinger engine in the bottom left of the picture.

Q: How large is the engine and how much does it weigh?

Roop: Overall, it's about 10 inches (because of the gear drive) and about 4 inches wide and about 4 inches tall [25 x 10 x 10 cm], and it weighs approximately a pound (0.5 kg).

Q: What other things might people be interested in about this engine?

Roop: People should know that it will put a big smile on their face if they build one and see it run. They should know that it's made from things that everybody either has, or can get at the hardware store. The power piston is made from a gas water pipe cap. It's just a piece of cast steel.

Q: If someone were starting out to build a new project like this, what kind of grill would you recommend that they choose?

Roop: I think it would be easiest to use any propane grill. I've looked at a bunch of them and they are all made basically the same. As long as it has a rotisserie. The bracket for most rotisserie motors sits on the outside of the barbecue and the rotisserie rod will typically run right through the center line of the lid and the lower body of the grill. I think the Chicken Flinger engine could be easily adapted to any barbecue grill.

Q: Where would someone want to use this kind of rotisserie?

Roop: I've taken it camping. It was really great because even though I typically try to get a camping location with electricity so I can have lights, this particular time, I knew that there wasn't going to be any electricity. I wanted to go fishing and not have to worry about cooking. So while the chicken was flinging on the rotisserie with no electricity required, I was fishing.

There are lots of places where people need to cook and they don't have electricity. A lot of back porches don't have an electrical outlet on the porch, so people have to run a cord out the door and it's easy to trip over it. That's what got me going on this project. I tripped over the cord that was running out of my house and down the steps. That particularly peeved me because I could have broken my neck.

In the next chapter you will learn what a technology has to do to clean up the world's ten most polluted cities.

Chapter 12
How to Clean up the World's Ten Most Polluted Cities

If our goal is to clean up the world's ten most polluted cities—and perhaps the rest of the world while we are at it—it would be worthwhile to learn how new ideas move through cultures. But before we dive into the topic that can help you predict which things will become a brief fad, which things will become modestly important, and which ones will go on to change the world, please take this short quiz.

Question. "If you build a better mousetrap the world will: (choose one)

a. beat a path to your door

b. make you rich and famous for your wonderful contributions to society

c. ignore you unless you do an awesome job of marketing your mousetrap

d. order you shot at sunrise for killing the sacred mice

e. none of the above.

No matter which answer you chose, I hope it is very clear to you that it takes more than great technology to change the world. As I write this, I have on my desk a brochure published by the US Department of Energy about Stirling engines in 1979. The last paragraph titled "Benefits" reads as follows.

"The projected returns of this national investment are enormous, amounting to billions of dollars over the next decades. Their benefits will be felt throughout our society and will show their greatest impact in conservation of energy."

Have you noticed the billions of dollars of benefits to United States society over the decades since 1979? Neither have I. However, the government did spend millions of dollars on Stirling engines, and new Stirling engine technology really was invented. Hopefully it's clear that it takes more than just a great technology to have any impact at all on society. But exactly what are the extra things that are needed for a technology to really catch on and help clean up the world? What is the "something extra" that comes between a great technology for making the world cleaner and the world actually becoming cleaner?

Imagine for a moment that you have developed a new technology that solves the world's power problems. Never mind what this technology really is (other than confirming that it's based on good physics). We will assume that it is a black box that makes cheap, clean, and reliable power. Inside this black box is a power source (it could be a Stirling engine, but for purposes of discussion it really doesn't matter) that can be used to generate electricity, power vehicles, and do almost anything that people need power for. To make this an even better idea, we will assume that the fuel that powers this box is 100 percent free.

Would the people who learned about your black box want to honor you, copy you, or stifle you? Would the news of your black box be front-page headline news that was flashed instantly all around the world? Or would you have a hard time getting reporters to return your phone calls?

If your black box was completely real and based on good physics, how long do you think it would take between the invention of this wonderful box and its widespread use around the world?

If you thought that it might take years or even decades for such a box to become widely used, you are right. If you thought it might never come into wide use, you would have plenty of historic precedent for that opinion as well. No matter how good a technology and no matter how promising it is, it will take from several years to several decades for it to become widely used.

Fortunately a new branch of science has been developed that studies how ideas move through a culture. It's called the science of memes, and it provides an uncommonly clear understanding of what an environmentally friendly technology such as Stirling engines has to be and do in order to catch on. With a proper understanding of memetics you can develop an almost uncanny skill at predicting what technologies will live or die in the world around you.

So what is a meme? A meme is an idea (any idea) that can be transmitted from human being to human being in the same way that a biological virus can be transmitted. If your neighbor sneezes on you, the biological viruses that land on your nose might give you a cold. But you don't realize that if your neighbor invites you to a party and sings "Happy Birthday" to the guest of honor, that you might just as quickly catch the idea, or the "meme," that singing "Happy Birthday" at a birthday party is a good thing to do.

People are uncommonly good at copying other people. This seems to be a unique human ability. While monkeys can swing effortlessly through trees, people can effortlessly see what other people have done and copy them. For all the talents of apes and other primates, apes do not "ape," but if you smile at a human baby, it will smile back.

In her excellent book on memetics titled *The Meme Machine* (Oxford University Press, 1999), Susan Blackmore says that for a meme to be successful, it needs to have three characteristics. It needs to have fecundity (or the ability to produce lots of offspring), fidelity (or the ability to make accurate copies of itself), and longevity (the ability for those copies to last for a very long time).

If we were to transfer those words to the human world of a society, we might say that cultures in which individuals have lots of sex, make lots of children, who look a lot like their parents and live a long time will thrive, because they have great memes. On the other hand, cultures such as that of the Shakers, the 19th-century religious organization in the United States that emphasized quality but required its members to be celibate, would be expected to die out, and indeed while antique Shaker chairs are still prized collectors items, the Shaker religion has become extinct.

A different American religion that was also founded in the nineteenth century at one time promoted polygamy, has always has promoted large families and almost immediately developed an amazingly effective program of evangelism. Today their membership numbers in the millions because they had great memes. The Shakers chose memes that led to the extinction of their religion, while a competitive American religion, formed about the same time, chose better memes and today has an annual operating budget larger than many small countries.

So how can memetics and an analysis of 19th-century American religions help us with our goal of cleaning up the world's ten most polluted cities? It's clear that the technology itself is only part of the solution. The technology also has to have great memes to carry it along.

So what are the necessary memes that a new technology must have in order to succeed? First it must have good fecundity. In other words it must be easy to copy and it must produce a lot of "offspring." The second criterion necessary for success would be good "fidelity." In other words, the copies must be accurate reproductions of the original. Finally the technology must have great "longevity." The machines that are built must last for a very long time.

So how have Stirling engines done in the meme race? The answer, of course, is that they have not done very well. For all his talents as a scientist and an engineer, Robert Stirling did not go into the business of building Stirling engines. He remained employed as a minister of the Church of Scotland until his death. He only built a few engines. Susan Blackmore would say that "the fecundity of early Stirling engines was not good." Robert Stirling also did not write a design guide for Stirling engines that would have made it easy for others to achieve the same excellent results that he did. So the fidelity of the early Stirling engines was also not good. The people who built Stirling engines in large numbers between the 1870s and the 1920s didn't succeed in building Stirling engines that were as good as the ones that Robert and James Stirling were building in the 1840s and 1850s.

Early Stirling engines were built of cast iron. Cast iron will oxidize through fairly quickly when exposed to a continuous flame in a Stirling engine, resulting in a failure after perhaps a year or so of operation. In other words, early Stirling engines also failed to have good "longevity."

So how can our understanding of memetics help us understand the outlook for Stirling engine technology today? Do modern Stirling engines have what it takes to succeed in the meme race? I think so. At least two companies covered elsewhere in this book are gearing up for high-volume production of Stirling engines to be used in the same role as furnaces or boilers in houses with the side benefit of producing electricity for use in the house. The potential for building millions of copies of these engines is very real. Susan Blackmore might say, "The outlook for fecundity looks good."

The ability of engineers to design and build Stirling engines has dramatically improved in the last ten years. Thanks to books by Allan Organ the concept of designing Stirling engines by dynamic similarity is now becoming better known. The publisher of this book, American Stirling Company, is actively looking for individuals or companies who have plans for power-producing Stirling engines that they would like to publish. So we can see that Stirling engine fidelity has dramatically improved.

When it comes to longevity, there is no other engine in the world today that can approach the longevity of a well-developed Stirling engine. Some Stirling engines have operated for years unattended and there are simply no other engines that compare with the reliability of modern well developed Stirling engines.

So what else is necessary for Stirling engines to win the meme race and become a part of our everyday lives? If you turn on your television set tonight and watch a movie about anything that has happened since 1900, it is extremely likely that the heroes in the movie will be using automobiles. While you don't think about it, every movie in which the heroes use cars is also a commercial for cars. It gets the idea of using cars in our heads. It puts the "car meme" in our minds.

I am one of those engineers who has often had the common engineering character flaw of thinking that every problem in the world has an engineering solution. But even outstanding engineering solutions and great engines will not necessarily make the world beat a path to your door. The engine has to have great memes. Those memes need to push hot human buttons and they also need to get in front of the largest audience possible. This brings us to the subject of Stirling engines in Hollywood movies.

It may surprise you to learn that there are some old movies that show Stirling engines. Since Stirling engines were widely used in the early part of the 20th century to power fans where electricity was not available, there are movies that show someone in an isolated location, using a Stirling-powered fan to stay cool. Unfortunately no one has been staying cool in a Hollywood movie with a Stirling-powered fan for over 50 years.

In order for a technology to win the memetics race, it has to become something that is on television. Perhaps we might discover that Hollywood scriptwriters are more powerful in pushing clean technology than are the world's best Stirling engine designers. If a blockbuster movie came out in which the heroes just happened to use clean Stirling engines, it would go a long way toward infecting people with the meme that the right way to live their lives is with clean-burning Stirling engine technology. Stirling engines have to move out of the labs of engineers like me and into the minds of scriptwriters who can include them casually in the script as just something that the heroes happen to do.

In the book *Virus of the Mind* (Integral Press, 1995), Richard Brodie points out that human beings have several hot buttons that can be pushed to promote whatever it is that you have in mind. Humans have a "sex" hot button, and so advertisers have learned that they should tie their products into sex in order to make them sell. We have a "help the children" hot button, and so charities have learned that they need to promote their worthy causes with children who are crying out for your help. We all have a "crisis" hot button and so politicians have learned that there should be an "energy crisis," a "health care crisis," and a "crisis of the day" in order to get elected.

So in addition to having great fecundity, fidelity, and longevity, a technology that really helps clean up the world's ten most polluted cities should also push the ten hottest human buttons. Do you find it hard to believe that Stirling engines could push people's "sex" button? I think that Madison Avenue advertising copy writers could probably figure out a way to get men to imagine opening a door and seeing the woman of their dreams lying on a sheepskin rug in front of the fireplace, and she's wearing nothing but a model MM-6 Stirling engine [see Figure 1-2]. Television commercial writers could probably figure out a way to make men think about two wonderful sources of Stirling engine power when they see a scantily clad woman.

And of course some advertising person who is much better than I am would be able to entice readers to go to their computers type in http://www.stirlingengine.com and order a MM-6 Stirling engine right now.

I know it's a silly example, but you have to admit that even a technology like Stirling engines could be marketed so that it would push the human sex hot button. Products that push human hot buttons win in the marketplace, and as much as I hate to admit it, good marketing beats good engineering every time. Other buttons that should be pushed include the "help the children" button and the "crisis" button. Tell people that Stirling engines will make the world cleaner for their children, and tell them that the world will be hit by a meteor, turning humans into the next dinosaurs if they don't buy a Stirling engine today, that would push the crisis hot button.

I want to be 100% clear that I am not encouraging you to lie, but even if you have a truly great technology, it still has to be sold. Of course when it comes to selling a new technology, all men are definitely *not* created equal. In the book *The Tipping Point* (Little Brown, 2001), Malcolm Gladwell tells the story familiar to all Americans of Paul Revere's ride. Paul Revere rode his horse through the Massachusetts countryside in 1775 to warn the American colonists that the British regular army was coming to seize arms and round up the colonial leadership. Paul Revere was spectacularly successful. He knew everyone along the route. He was a member of every club and social organization in Boston, and when it came time to get the word out, he really knew how to do it.

What most people don't know is that a different revolutionary, a tanner by the name of William Dawes, set out on the same urgent errand to deliver the news to the towns west of Boston. But because Dawes was not the same kind of "connector" that Paul Revere was, he didn't have the same impact. His message didn't "tip." So in our goal of cleaning up the world's ten most polluted cities, we know that we would have to enlist some of the world's greatest connectors and the world's greatest salesmen.

So what does it take for a technology to tip? What does it take for something new to get to the place where everyone in an entire culture uses it? Memetics would say that it takes a technology with great fecundity, fidelity, and longevity. The tipping point theory would say that it takes people like Paul Revere, who can move an idea quickly through a culture. The idea has to be pushed, and it has to be sold. It has to be something so exciting that it can sizzle in people's imaginations. It has to be something that can get great salesmen involved. It has to be something that can move the world.

Can Stirling engines do that? Of course they can, but only if they remember to choose great memes.

Chapter 13
The Top Five Modern Uses for Stirling Engines

A vast portion of the effort invested in Stirling engines over the past many years has been wasted. It's been wasted because people have worked hard to develop Stirling engines for applications for which they are not well suited. So where should the Stirling engine development efforts of policymakers and businessmen, engineers and salesmen be directed today?

Fortunately there are several important markets for which Stirling engines are simply the best engine in the world. Here are several examples.

Stirling engines should be used today for any application in which people are willing to pay a premium either for silence or extremely low vibration. The application that comes to mind is spacecraft. Deep-space probes and planetary landers often carry extremely sensitive instruments and cameras that cannot operate in a high-vibration environment. Since Stirling engines do not have internal explosions, they can be built to be extremely smooth-running, 100 percent mass-balanced machines.

Another application for which people are willing to pay a premium for silence is to power submarines. Military submarines depend on silence for their effectiveness. Also, research submarines would prefer not to startle marine life with a noisy engine; and luxury submarines can be built that use Stirling engines for their extended underwater operational capabilities. Other military applications would include any mission where silence is a requirement.

The second major category for using Stirling engines today would be any application in which the per-unit costs are very high and the competing technology is inadequate. As strange as it may seem, power plants for propeller-driven, piston-powered airplanes have not changed much since the 1950s. The best 200-horsepower aircraft engine that you can buy costs about the same amount as a new Cadillac, and is remarkably less reliable than the engine in the cheapest new car. This could be an excellent market in which to introduce a Stirling engine.

Another excellent use for Stirling engines that falls under the same category would be auxiliary power for sailboats and yachts. Some sailors enjoy sailing because they like it quiet and they don't want to have to fire up a diesel engine to come back into the harbor. Any marine application of the Stirling engine is significantly helped by the fact that good cooling water is available.

A third item under this category would be "indoor power-producing engines." There are many products that range from electric wheelchairs to golf carts that need more power than can be conveniently stored in batteries and also need to operate indoors. There is clearly a market for an auxiliary generator that produces an exhaust clean enough not to be a problem indoors. Internal combustion engines, no matter how well refined, are extremely unlikely to meet this standard of clean operation.

The third major category for using Stirling engines today is any place where multi-fuel use is extremely important. Does the engine need to be fueled by solar energy on sunny days and by natural gas at night and on days when the sun is not shining? While there are a few other engines, such as gas turbines, that can operate on several different refined fuels, there is nothing that can match a Stirling engine's versatility for simplicity and multi-fuel use. Gas turbines can be made to run on gasoline, jet fuel, or natural gas, but you won't get much power out of a gas turbine engine if you try to run it on sunshine.

Solar cells are wonderful solid-state devices, but they don't deliver much power at night, when a solar Stirling engine could still be running on clean-burning natural gas.

The fourth top category for using Stirling engines today is fixed power generation and heating for either remote or city use. First let's talk about the remote power generation. If you have a beautiful cabin on a lake in Alaska or northern Canada, you may be surrounded by abundant supplies of biomass fuel in the form of wood, but bringing in refined fuel such as gasoline or diesel to run a conventional generator might be a significant drawback. Stirling engines can be built to run on any heat source, including the free wood that you might have available.

Stirling engines can also be made to run almost indefinitely, making them the perfect power source for remote research and weather stations. Some Stirling engines have been built that were powered by radio isotopes. After many years of careful development, the results were engines that would run almost indefinitely. While not every Stirling engine is an extremely reliable machine, Stirling engines indeed have been built that will run almost indefinitely.

People who live in cities and the suburbs have a need for both heat and power. Stirling engines can excel in this area when they are used to replace the heating system in a house while producing electricity essentially as a bonus.

When a power-producing company decides that it needs to add capacity, the fuel of choice tends to be natural gas. But when this gas is burned at a remote generating facility 100 miles from the customer, a tremendous amount of the natural gas is wasted long before it reaches your house. Stirling engines can, and should, be integrated into furnaces and home heating systems. These Stirling heaters would provide reliable heating even when the electricity goes off.

Portions of the United States, from the Great Plains to the Northeast, are subject to severe ice storms. These ice storms often knock down the power lines for weeks, causing tremendous disruption and many deaths. It's worth noting that natural gas supplies delivered through underground pipes are rarely disrupted during ice storms. What usually happens is that the electric supply is cut off, rendering the furnace totally worthless, even if there is propane in a tank behind the house or if there is natural gas in the pipes.

The fifth top category for using Stirling engines today is for waste heat recovery in any application where the waste-heat stream temperature is over 650° C (1,200° F). If the waste-heat source produces heat twenty-four hours a day, seven days a week, then the application is even more compelling. The glass-manufacturing industry fits this profile perfectly. Glass manufacturing produces extremely hot, very clean waste heat day and night with some installations running for a year or more between shuts downs.

Any other industrial application where similar conditions apply is a candidate for Stirling engine use. Any industry that exhausts a large amount of high-temperature waste heat may be able to save millions of dollars with the right industrial-scale Stirling engines. (As this book goes to press these engines are still in the development phase.)

It's worth discussing the practicality of using Stirling engines for low-temperature waste heat recovery. The Stirling engines that run on tiny temperature differences, manufactured by the American Stirling Company, [see Fig. 1-1 And 1-2] often prompt our customers to call us with requests to for power-producing engines in the tens to hundreds of kilowatts range that would be powered by low temperature waste heat sources from 120 to 500° F (48 to 260° C)

Theoretically these engines could be built from low-cost plastics and inexpensive aluminum alloys. From Boston to Bombay the world is full of millions of low-temperature waste heat sources. Unfortunately the drawback is usually that Stirling engines tend to get unacceptably big as the temperature differences on which they run become smaller. The models made by American Stirling Company are delightful educational tools and toys, but they produce extremely tiny amounts of power.

Could practical Stirling engines be built that would run on small temperature differences? Unfortunately, the answer is usually no. A Stirling engine that would produce 10 kW (13.4 hp) of power on the temperature difference between boiling water and room temperature would probably be an excessively large engine. This should not discourage future experimenters from developing such an engine if they feel that they have a good application.

However it's good to put future developers of low temperature difference power producing engines that they should expect to build very large engines. It may be possible to make very big low temperature difference engines work in some situations. Perhaps a Stirling engine could be integrated into a metal roof design, making a building that would not appear to have any solar collectors at all but would produce power from a collector built into the roof.

When you are thinking about low temperature difference Stirling engines, it's worthwhile to compare the production of mechanical power to electrical power. To produce electrical power, you need volts times amps to make watts. To produce mechanical power with a Stirling engine, you need temperature (the approximate equivalent of volts) times heat (the approximate equivalent of amps) to yield power. In other words, you need both temperature (quality) and heat (quantity) to make power. Low temperature difference Stirling engines need to move a massive amount of heat at a low temperature to make power.

Should Stirling engines ever be used in automobiles? The answer is a firm no, with one modest exception. Stirling engines do not make good engines in automobiles for reasons discussed in other chapters. (They are slow-starting engines, and regulating them is more of a challenge than regulating internal combustion engines.) But Stirling engines clearly could be used in hybrid vehicles in which the batteries provide instant starting and the Stirling engine tops off the batteries when the charge gets low.

Unfortunately, it's extremely unlikely that Stirling engines will ever reach wide-scale use in this application for the following reason. Put yourself in the position of a lead engineer at a major auto company. Your management has just given you the fascinating and wonderful job of building a hybrid vehicle using battery power plus a gasoline-burning engine to provide power. You can choose between using a fifth-generation Stirling engine and using a 100th-generation internal combustion engine. Multi-fuel use is not important because the engine will be powered by gasoline. Quiet running is not important because a muffler can make an internal combustion engine quiet enough for use in a car. The fifth-generation Stirling engine will require some significant development work, but the 100th-generation gasoline engine is almost sure to work right the first time. Which one will you choose? Obviously, you're going to make the choice that will make your job easier.

But there ARE thousands of other places in which the use of the Stirling engine will make your life (and your job) dramatically easier because competing power sources have unacceptable drawbacks. A wonderful engine in an application for which it is not well suited makes a bad overall product. But the same wonderful engine running in an application to which it is well suited, really should be called, "The Perfect Engine."

From outer space to silent submarines, from cordless hair dryers to warm, well-lit houses in the suburbs, Stirling engines are ready to fill a vital role in the world's never-ending quest for power.

The last chapter is a re-print from an article that I wrote for the three volume library reference book titled *"Macmillan Encyclopedia of Energy."* Much of the information in chapter 14 is covered elsewhere in this book.

Stirling Engines

Re-printed by permission from
Macmillan Encyclopedia of Energy
by Brent H. Van Arsdell

The principle that makes Stirling engines possible is quite simple. When air is heated it expands, and when it is cooled it contracts. Stirling engines work by cyclically heating and cooling air (or perhaps another gas such as helium) inside a leak tight container and using the pressure changes to drive a piston. The heating and cooling process works like this: One part of the engine is kept hot while another part is kept cold. A mechanism then moves the air back and forth between the hot side and the cold side. When the air is moved to the hot side, it expands and pushes up on the piston, and when the air is moved back to the cold side, it contracts and pulls down on the piston.

While Stirling engines are conceptually quite simple, understanding how any particular engine design works is often quite difficult because there are hundreds of different mechanical configurations that can achieve the Stirling cycle. Figure 14-1 shows a schematic of a transparent educational demonstration engine that runs on the top of a cup of hot coffee. This engine uses a piece of foam similar to what would be used as a filter for a window air conditioning unit to "displace" the air between the hot side and the cold side. This foam displacer is carefully mounted so it does not touch the walls of the cylinder. Figure 14-2 shows how this particular engine achieves the Stirling cycle. In this engine, the air flows through and around the displacer from the hot side then back to the cold side, producing a power pulse during both the hot and cold portion of the cycle. Stirling engines can be mechanically quite simple since they have no valves, and no sparkplugs. This can result in extremely high reliability as there are fewer parts to fail.

It is worthwhile to compare Stirling engines to other more familiar engines and note their similarities as well as their differences. Stirling engines are a type of heat engine. They turn heat into mechanical work and in this sense they perform the same function as other well known heat engines such as gasoline, diesel, and steam engines. Like steam engines, Stirling engines are external combustion engines, since the heat is supplied to the engine from a source outside the cylinder instead of being supplied by a fuel burning inside the cylinder. Because the heat in a Stirling engine comes from outside of the engine, Stirling engines can be designed that will run on any heat source from fossil fuel heat, to geo-thermal heat, to sunshine. Unlike steam engines, Stirling engines do not use a boiler that might explode if not carefully monitored.

When operating on sunshine, or geo-thermal heat, Stirling engines obviously produce no pollution at all, but they can be exceedingly low emissions engines even when burning gasoline, diesel, or home heating oil. Unlike gasoline or diesel engines that have many thousands of start stop cycles of combustion each minute, burners in Stirling engines burn fuel continuously. It's much easier to make a continuous combustion engine burn very cleanly than one that has to start and stop. An excellent demonstration of this principle is to strike a match, let it burn for a few seconds, then blow it out. Most of the smoke is produced during the starting and stopping phases of combustion.

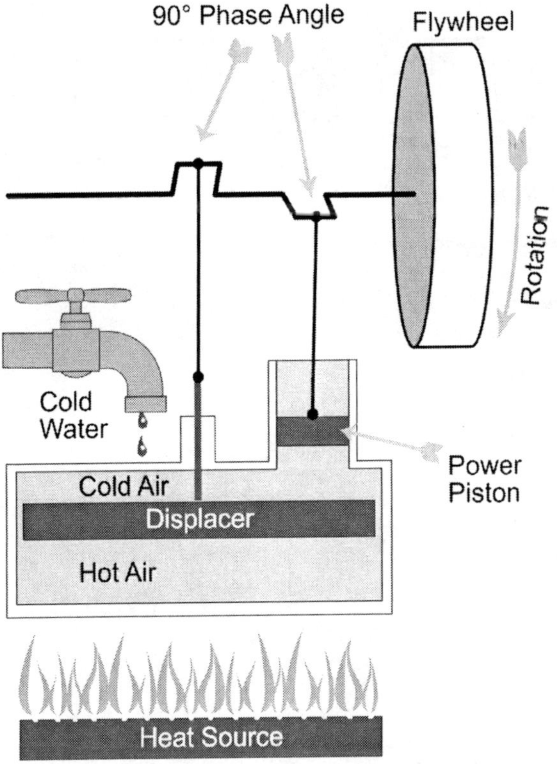

Fig. 14-1: A schematic of a transparent educational demonstration engine that runs on the top of a cup of hot coffee.

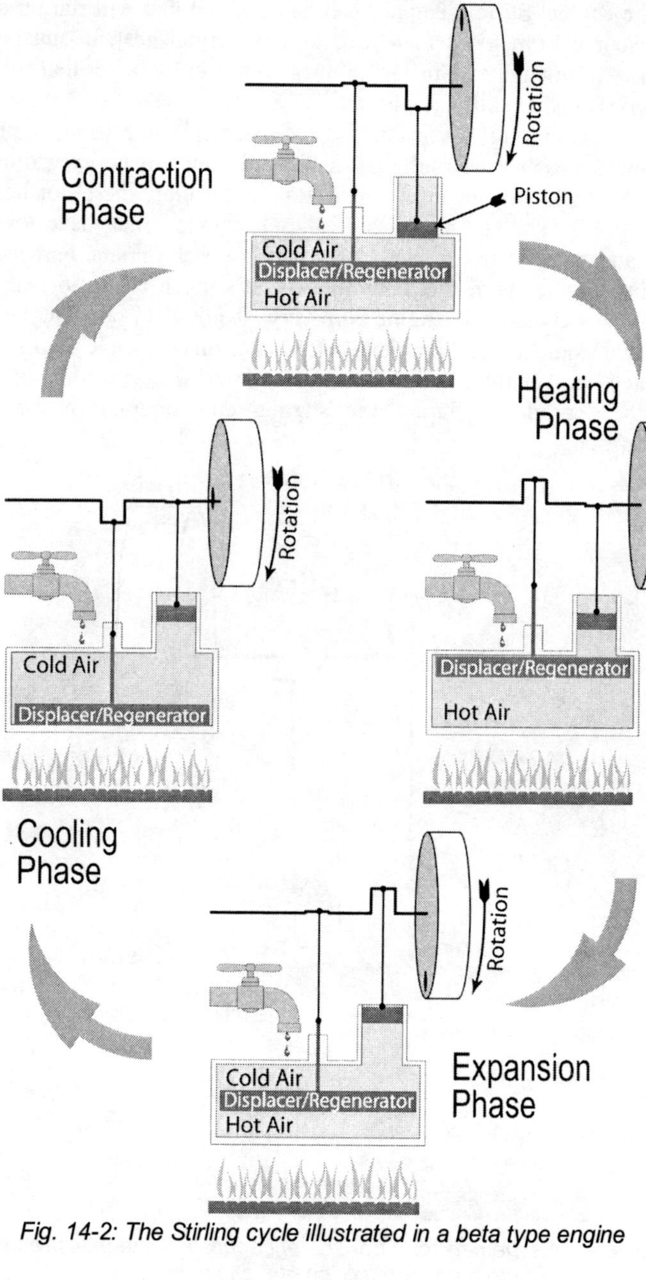

Fig. 14-2: The Stirling cycle illustrated in a beta type engine

A Brief History

In the early days of the industrial revolution, steam engine explosions were a real problem. Metal fatigue was not well understood, and the steam engines of the day would often explode, killing and injuring people nearby. In 1816 the Reverend Robert Stirling, a minister of the Church of Scotland, invented what he called "A New Type of Hot Air Engine with Economiser" as a safe and economical alternative to steam. His engines couldn't explode, used less fuel, and put out more power than the steam engines of the day.

The engines designed by Robert Stirling and those who followed him were very innovative engines, but there was a problem with the material that was used to build them. In a Stirling engine, the hot side of the engine heats up to the average temperature of the flame used to heat it and remains at that temperature. There is no time for the cylinder head to cool off briefly between power pulses. When Robert Stirling built his first engines, cast iron was the only readily available material, and when the hot side of a cast iron Stirling engine was heated to almost red hot, it would oxidize fairly quickly. The result was that quite often a hole would burn through the hot side causing the engine to quit. In spite of the difficulties with materials, tens of thousands of Stirling engines were used to power water pumps, run small machines, and turn fans, from the time of their invention up until about 1915.

As electricity became more widely available in the early 1900s, and as gasoline became readily available as a fuel for automobiles, electric motors and gasoline engines began to replace Stirling engines.

Regeneration

Robert Stirling's most important invention was probably a feature of his engines that he called an "economiser." Stirling realized that heat engines usually get their power from the force of an expanding gas which pushes up on a piston. The steam engines that he observed dumped all of their waste heat into the environment through their exhaust and the heat was lost forever. Stirling engines changed all that. Robert Stirling invented what he called an "economiser" that saved some heat from one cycle and used it again to pre-heat the air for the next cycle.

It worked like this: After the hot air had expanded and pushed the piston as far as the connecting rod would allow, the air still had quite a bit of heat energy left in it. Stirling's engines stored some of this waste heat by making the air flow through economiser tubes that absorbed some of the heat from the air. This pre-cooled air was then moved to the cold part of the engine where it cooled very quickly and as it cooled it contracted, pulling down on the piston. Next the air was mechanically moved back through the pre-heating economiser tubes to the hot side of the engine where it was heated even further, expanding and pushing up on the piston. This type of heat storage is used in many industrial processes and today is called "regeneration." Stirling engines do not have to have regenerators to work, but well designed engines will run faster and put out more power if they have a regenerator.

Continued Interest

In spite of the fact that the world offers many competing sources of power there are some very good reasons why interest in Stirling engines has remained strong among scientists, engineers, and public policy makers. Stirling engines can be made to run on any heat source. Every imaginable heat source from fossil fuel heat to solar energy heat can and has been used to power a Stirling engine.

Stirling engines also have the maximum theoretical possible efficiency since their power cycle (their theoretical pressure volume diagram) matches the Carnot cycle. The Carnot cycle, first described by the French physicist Sadi Carnot, determines the maximum theoretical efficiency of any heat engine operating between a hot and a cold reservoir. The Carnot efficiency formula is: (T(hot)-T(cold))/T(hot). T(hot) is the temperature on the hot side of the engine. T(cold) is the temperature on the cold side of the engine. These temperatures must be measured in absolute degrees (Kelvin or Rankine).

Stirling Applications

Stirling engines make sense in applications that take advantage of their best features while avoiding their drawbacks. Unfortunately, there have been some extremely dedicated research efforts that apparently overlooked the critical importance of matching the right technology to the right application.

In the 1970s and 1980s a huge amount of research was done on Stirling engines for automobiles by companies such as General Motors, Ford, and Philips Electronics. The difficulty was that Stirling engines have several intrinsic characteristics that make building a good automobile Stirling engine quite difficult. Stirling engines like to run at a constant power setting, which is perfect for pumping water, but is a real challenge for the stop and go driving of an automobile.

Automobile engines need to be able to change power levels very quickly as a driver accelerates from a stop to highway speed. It is easy to design a Stirling engine power control mechanism that will change power levels efficiently, by simply turning up or down the burner. But this is a relatively slow method of changing power levels and probably is not a good way to add the power necessary to accelerate across an intersection. It's also easy to design a simple Stirling engine control device that can change power levels quickly but allows the engine to continue to consume fuel at the full power rate even while producing low amounts of power. However it seems to be quite difficult to design a power control mechanism that can change power levels both quickly and efficiently. A few research Stirling engines have done this, but they all used very complex mechanical methods for achieving their goal.

Stirling engines do not develop power immediately after the heat source is turned on. It can take a minute or longer for the hot side of the engine to get up to operating temperature and make full power available. Automobile drivers are used to having full power available almost instantly after they start their engines.

In spite of these difficulties, there are some automobile Stirling applications that make sense. Hybrid electric cars which include both batteries and a Stirling engine generator would probably be an extremely effective power system. The batteries would give the car the instant acceleration that drivers are used to, while a silent and clean running Stirling engine would give drivers the freedom to make long trips away from battery charging stations. On long trips, the hybrid car could burn either gasoline or diesel, depending on which fuel was cheaper.

To generate electricity for homes and businesses, research Stirling generators fueled by either solar energy or natural gas have been tested. They run on Solar power when the sun is shining and automatically convert to clean burning natural gas at night or when the weather is cloudy.

There are no explosions inside Stirling engines, so they can be designed to be extremely quiet. The Swedish defense contractor Kockums has produced Stirling engine powered submarines for the Swedish navy that are said to be the quietest submarines in the world.

Aircraft engines operate in an environment that gets increasingly colder as the aircraft climbs to altitude, so Stirling aircraft engines, unlike any other type of aircraft engine may derive some performance benefit from climbing to altitude. The communities near airports would benefit from the extremely quiet operation that is possible. Stirling engines make sense where these conditions are met:

1. There is a premium on quiet.
2. There is a very good cooling source available.
3. Relatively slow revolutions are desired.
4. Multiple fuel capacity is desired.
5. The engine can run at a constant power output.

6. The engine does not need to change power levels quickly.
7. A warm-up period of several minutes is acceptable.

Low Temperature Difference Engines

In 1983, Ivo Kolin, a professor at the University of Zagreb in Croatia, demonstrated the first Stirling engine that would run on a heat source cooler than boiling water. After he published his work, James Senft, a mathematics professor at the University of Wisconsin, River Falls built improved engines that would run on increasingly small temperature differences, culminating in an elegant and delicate Stirling engine that would run on a temperature difference smaller than 1° F.

These delicate engines provide value as educational tools, but they immediately inspire curiosity into the possibility of generating power from one of the many sources of low temperature waste heat (less than 100° C.) that are available. [See also co-generation] A quick look at the Carnot formula shows that an engine operating with a hot side at 100° C. and a Cold Side at 23° C. will have a maximum Carnot efficiency of [((373° K-296° K)/373 K) x 100] about 21 percent. If an engine could be built that achieved 25 percent of the possible 21 percent Carnot efficiency it would have about 5 percent overall Carnot efficiency.

That figure seems quite low until one realizes that calculating Carnot efficiency for an engine that uses a free heat source might not make much sense. For this type of engine it would probably be more worthwhile to first consider what types of engines can be built, then use dollars per watt as the appropriate figure of merit.

Stirling engines that run on low temperature differences tend to be rather large for the amount of power they put out. However this may not be a significant drawback since these engines can be largely manufactured from lightweight and cheap materials such as plastics. These engines could be used for applications such as irrigation and remote water pumping.

Cryocoolers

It isn't immediately obvious, but Stirling engines are a reversible device. If one end is heated while the other end is cooled, they will produce mechanical work. But if mechanical work is input into the engine by connecting an electric motor to the power output shaft, one end will get hot and the other end will get cold. In a correctly designed Stirling cooler, the cold end will get extremely cold. Stirling coolers have been built for research use that will cool to below 10 K. Cigarette pack sized Stirling coolers have been produced in large numbers for cooling infrared chips down to 80 K. These micro Stirling coolers have been used in high end night vision devices, antiaircraft missile tracking systems, and even some satellite infrared cameras.

Bibliography

Senft, James R. (1993). *An Introduction to Stirling Engines*. River Falls, Wisconsin: Moriya Press.

Senft, James R. (1996). *An Introduction to Low Temperature Differential Stirling Engines*. River Falls, Wisconsin: Moriya Press.

Walker, Graham. (1980). *Stirling Engines*. Oxford: Oxford University Press.

Walker, Graham.; Fauvel, Owen R.; Reader, Graham.; Bingham, Edward R. (1994). *The Stirling Alternative*. Yverdon, Switzerland: Gordon and Breach Science Publishers.

West, Colin. (1986). *Principles and Applications of Stirling Engines*. New York: Van Nostrand Reinhold.

Contact Pages:

You can contact many of the companies and individuals mentioned in this book at the addresses below.

Brent H. Van Arsdell - President
American Stirling Company
2726 Shelter Island Dr. # 172
San Diego, CA 92106
Phone 760-742-2727
Fax 858-777-3459
www.stirlingengine.com
email brent@stirlingengine.com

As we go to press our phone number was about to change. You may want to check our Web site for a more current phone number.

Kockums AB
SE-205 55 Malmö, Sweden

Phone +46 40 34 80 00
Fax + 46 40 97 32 81
email: information@kockums.se
www.kockums.se

U.S. SUBMARINES, INC.,
936 N.W. First Street
Ft. Lauderdale, FL 33311

Tel: 954/467-9028
Fax: 954/467-9584
Email: ussubs@ussubs.com
www.ussubs.com

Whisper Tech Ltd.
224 Armagh Street
Christchurch, New Zealand

Postal address:
PO Box 13705, Armagh
Christchurch 8031, New Zealand

Phone: +64 (3) 363 9293
Fax: +64 (3) 363 9294
info@whispertech.co.nz
www.whispertech.co.nz

STM Power, Inc.
275 Metty Drive
Ann Arbor, MI 48103 USA

Phone: 734-995-1755
Fax: 734-995-0610
www.stmpower.com

Allan J. Organ, Ph.D.
32 Pettit's Lane, Dry Drayton
Cambridge CB3 8BT United Kingdom

Phone + (0)1954 78185
Fax: +(0) 1954 781253
Email: allan_j_o@btopenworld.com

Stirling Technology Company
4208 West Clearwater Ave
Kennewick WA 99336-2626, USA

Phone: 509-735-4700 Ext. 112
Fax: 509-736-3660
Email: info@stirlingtech.com
www.stirlingtech.com

L. Merrick Lockwood, Ph.D.
Shelton Cottage, Lower Shola Road
Kodaikanal, TN 624101, India

Phone: (+91 4542) 241411
Email: merricklockwood@yahoo.com

Jim Symanski
P.O. Box 14
Garden Valley, CA 95633
Phone: (530) 263-5176
A good time to call is 7:00 p.m. Pacific Time

Craig Roop
4615 South 29th W. Avenue
Tulsa, OK 74107
Phone: (918) 446-9976
Email: volvonut2001@yahoo.com

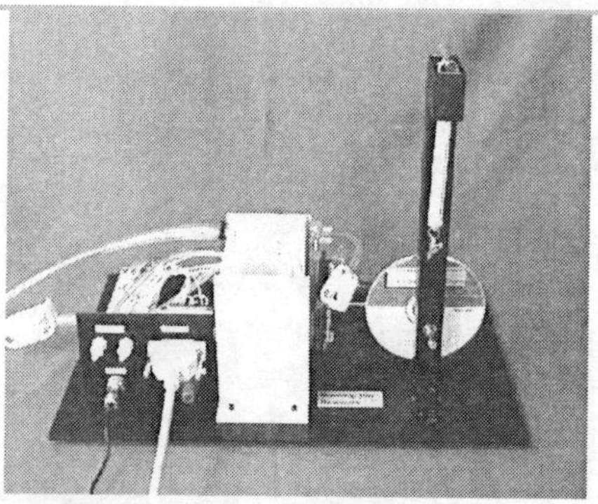

The Smart Stirling lab engine was developed at the request of University of Kentucky thermodynamics professor Jim Smart.

This engine is intended for physics and thermodynamics professors who want to provide a high-quality lab experience for their students. It comes with pressure, temperature, and volume sensors along with custom electronics to interface with a PC.

Please contact us for shipping details.

Our Price: $4,950.00

A limited edition Stirling engine with an elegant Victorian look.

The Van Arsdell Stirling engine has the elegant look of an antique Victorian engine. No expense has been spared in building this engine. All the brass components are machined from solid brass stock (nothing is brass plated), the parts are polished by hand, and the engine is carefully mounted on a clear-grain Honduran mahogany base.

This engine measures 11" long by 8" wide and stands 7 3/4" tall (~28 x 20 x 19.7 cm). All engines are engraved with a serial number.

The beautiful look of this engine was designed by Michael Crumpton; and Brent H. Van Arsdell did the concept and built a proof of concept prototype.

Shipping to the USA $35.00 Elsewhere $89.00

Our Price: $1,479.00

MM-6 Stirling engine

The amazing MM-6 Stirling engine will run indefinitely on just the heat from your warm hand! As a gift, conversation piece, or classroom demo, you can be sure it will be unique and attention getting.

This Stirling engine is a beautiful conversation piece for your home or office or a great physics or chemistry demo for your classroom or lab and everything in between. It will run indefinitely on your warm hands in a 72° F (23° C) room or on a computer monitor, TV, DVD player, VCR, stereo amplifier/receiver, fax machine, or even bright sunlight shining through a window. The possibilities are endless!

While the MM-6 operates based on well-understood physics, it may be the closest thing to magic and a perpetual motion machine you can buy anywhere at any price. Amaze your friends, your family, your co-workers, or just yourself! The MM-6 is a great unique present for that special friend, relative, business associate, or potential client who already has all the usual, mundane worldly possessions. You can be confident that he or she will not have an MM-6.

Our Price: $359.00 Shipping to the USA $17.00 Elsewhere $34.00

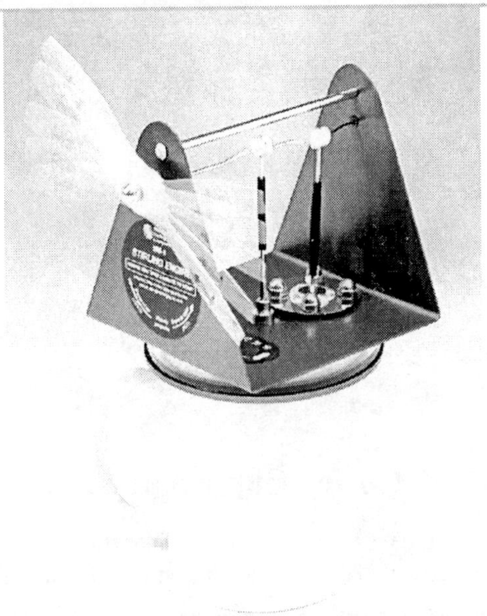

MM-1 Coffee Cup Stirling Engine

Our original transparent Coffee Cup Stirling engine.

Power this engine with a cup of steaming hot coffee or set it on a plate of ice chips to see it run in the opposite direction. The MM-1 has been used for educational demonstrations from middle-school science classes to Harvard thermodynamics classes.

This engine is also a favorite of engineers and techies all over the world! If you need to give a gift to someone who loves technology this is guaranteed to make them happy. The MM-1 comes ready to run and spins about 250 rpm on hot coffee or 100 rpm running on ice in a warm room.

Our Price: $179.00 Shipping to the USA $12.00 Elsewhere $16.00

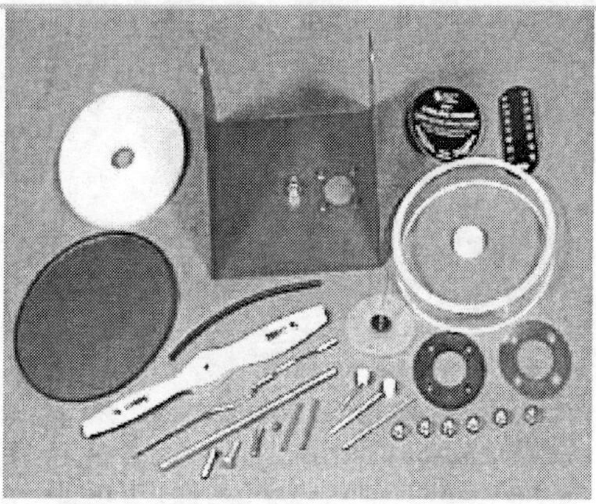

MM-5 Coffee Cup Engine Kit

This kit includes all the parts to build our original transparent engine that sells ready to run for $139.00 Read our online instructions.

Ready to assemble. You only need to purchase superglue and silicone sealant. You will also need a few common hand tools to assemble your engine. It's a pleasant two or three evening kit and a fun way to learn about Stirling engines.

You may also want to buy a spare diaphragm. This is the only wear part in the engine.

Our Price: $119.00 Shipping to the USA $12.00 Elsewhere $16.00

**Around the World by Stirling Engine:
Environmentally friendly Stirling engines,
their applications worldwide
and into space
by Brent H. Van Arsdell**

"How do Stirling engines work, and where can I get one that puts out some real power?" This book answers those questions. It also covers all the current power-producing uses of Stirling engines along with how to contact the people who build them.

You can expect a chapter on Stirling-powered submarines, Stirling auxiliary generators on yachts, perhaps even Stirling engines on deep space probes. If space probes aren't your thing you may enjoy reading about the man whose Stirling engine powers his workshop.

Bulk discounts available.

Our Price: $17.95 Shipping USA $5.00 Elsewhere $9.00

American Stirling Order Form

		Quantity	Subtotal
MM-1 Coffee Cup Stirling Engine $179.00	Shipping to the USA $12.00 Elsewhere $16.00		
MM-5 Coffee Cup Stirling Engine Kit $119.00	Shipping to the USA $12.00 Elsewhere $16.00		
MM-6 Stirling Engine $359.00	Shipping to the USA $17.00 Elsewhere $34.00		
Victorian Look Stirling Engine $1479.00	Shipping to the USA $35.00 Elsewhere $89.00		
Smart Stirling Lab Engine $4950.00	Shipping to the USA $65.00 Elsewhere $350.00		
Around the World by Stirling Engine Book $17.95	Shipping to the USA $5.00 Elsewhere $9.00		
	Total		

American Stirling Company
2726 Shelter Island Dr. # 172
San Diego, CA 92106
Phone 760-742-2727 • Fax 858-777-3459
www.stirlingengine.com
email: brent@stirlingengine.com

Order Form